WILDBORE

This book is dedicated to my daughter Caitlyn and son Carter.

WILDBORE
A photographic legacy

Catherine Knight

TOTARA
PRESS

First published in 2018 by
Tōtara Press
Pohangina Valley
NEW ZEALAND
www.khm.co.nz

ISBN 978-0-473-44413-6

A catalogue record for this book is available from the National Library of New Zealand.

Cover and book design and layout: Anthony Behrens, www.swampthing.co.nz
Map: Tim Nolan, Blackant Mapping Solutions
Editor: Helen Eastwood

Printed in Masterton, New Zealand by Greenlees Print.

Published with assistance from the Earle Creativity and Development Trust.

Contents

A bush-clearers' camp in Komako, Pohangina Valley, circa 1904. The gang of 'bush-whackers' were employed by Dick Port and Alex Stewart as part of their bush-felling contract on H.J. Rutherford's farm. Dick Port is standing centre (with right hand on hip and left hand holding an axe). The boy sitting in the foreground (with axe on knee) is Edgar Savell.

C.E. Wildbore, Palmerston North Library

Foreword

For many years I have made use of Charles Wildbore's photographs in my teaching and, in a couple of instances, to illustrate publications. When I first moved to Palmerston North in the early 1980s to take up a post at Massey University, I was taken on a drive to visit Totara Reserve and other sites in the Pohangina Valley. Not long afterwards, I came across some of Wildbore's evocative photographs of timber fellers at work, the bush burn and the sowing of grass seed by hand on steep and recently burnt slopes. At the time, these photographs were mounted on large pieces of heavy cardboard and stored in cabinets in the public library. The Pohangina Valley, as a site of very recent European farm settlement, had featured in a number of Massey University history and geography theses, most notably Lynette Wright's 1968 Master's thesis, '"The axe bites deep": Settlement and land use in the Pohangina County, 1863–1963'. For me, the contrast between the Pohangina District and the landscapes I was more familiar with – Canterbury and the agricultural history of the 'great estates', centred on the production of wool and wheat – was marked. The rapidity and extensiveness of environmental change in the valley, a process representative of the clearance of much of the forest of Te Ika-a-Māui/North Island, was something I learned of only later.

It is with particular pleasure, therefore, that I write this foreword. The Pohangina Valley has featured occasionally and in a minor way in my research over several decades. The book brings together a remarkable photographic record that is of importance to historians and environmental historians, both locally and nationally, as well as to photographic historians and to a broader general audience. It also brings the creator of the photographs into the field of vision. The person behind the camera is typically invisible, albeit that Wildbore's photograph 'the travelling tinker' is revealed as a carefully constructed arrangement of his children and himself. It adds to the value of the photographs to have this account of Wildbore's life and the times in which he lived. Indeed, it is possible to argue that the photographs themselves also add in ways that words cannot to an understanding of Wildbore's life as a pioneer settler in the Pohangina Valley.

Professor Michael Roche
School of People, Environment and Planning
Massey University
April 2018

Part 1

A TIME OF UNPRECEDENTED CHANGE

Many people will know Charles Wildbore's photographs even if they know nothing about the man who took them. His photographs of the Pohangina Valley[1] around the turn of the 20th century are found adorning the walls of Manawatū cafés and libraries, and his images of timber-milling, bush-'whacking' and newly hewn farmscapes have featured on the covers of and within books documenting New Zealand's history.[2] His turn-of-the-century series showing the transformation of near-primeval forest country to pastoral farms is perhaps the only example of this process being captured in its entirety in New Zealand. Yet, despite the importance of his photographs to our understanding of our country's environmental past, little has been

The Moar farm on Pohangina Road, looking south towards Pohangina Village, circa 1910. The two boys sitting on the fence in front of the cowshed are likely to be (from left) brothers Norman and Herbert (Bert) Moar. The child standing to their left may be Pearly (their younger sister). The bearded man is likely to be their father, Simon Moar, who emigrated to New Zealand from the Orkney Islands in 1878. The woolshed can be seen centre-right, and the Moar home beyond that, on the right.

C.E. Wildbore, Palmerston North Library

written about Charles Edward Wildbore (1861–1937) himself – a farmer, beekeeper, inventor and photographer, who, with his fast-expanding family, settled in the Pohangina Valley in the 1890s.

Wildbore's photographs are the unique product of place, time and agency. The place was the Pohangina Valley, and the time was the last decade of the 19th century and the first decades of the 20th century. The agency was Wildbore – a naturally inquisitive and creative man who witnessed this place at this time, and felt compelled to photograph it. He captured the transformation of this comparatively remote, forest-covered valley in the Manawatū region from a forested wilderness, through the phases of forest-milling, bush burning and the sowing of seed amid charred tree stumps and limbs, to productive farmland – a process that had repeatedly unfolded throughout much of the North Island by this time. In this way, therefore, Wildbore was witness to the dramatic landscape transformation that had already reshaped much of New Zealand.

What made Wildbore remarkable was that he had the foresight to capture this process of transformation, and the everyday activities that followed, through his photography. His photographs have helped us visualise a New Zealand of the past – a landscape so different to the one we see today that we would otherwise struggle to imagine it. His landscape photographs are significant, not just in a regional context, but in the national context also. Many of them capture New Zealand in a phase of transition from 'wilderness' to 'civilisation' – a consequence of environmental change that was unprecedented in both its pace and scale.

Charles Wildbore in the drawing room of an unidentified home (not his own) circa 1910.

C.E. Wildbore, Palmerston North Library

Capturing the 'final frontier' of environmental transformation

In the late 19th and early 20th centuries, the Pohangina Valley could be described as a final frontier – a forested hinterland not yet shaped into the pastoral farmland that, by this time, stretched across the lowlands and hill country of New Zealand. While other regions had been stripped of their forest and enveloped in a grass sward decades before, the Pohangina Valley was only just entering this phase in its development – a consequence of its ruggedness and its poor accessibility.[3] The significance of the history of the Pohangina Valley can therefore only be fully understood by appreciating the preceding settlement history of the rest of New Zealand.

By the second half of the 19th century, much of the open country in regions such as Canterbury, Otago and Hawke's Bay had been settled, often in large estates or sheep runs that were well beyond the means of the working-class man.[4] From the 1870s, the government turned its focus to the provision of land for the settlement of those of lesser means; the so-called 'small man'. It introduced legislation that encouraged the settlement of land by people who did not have the funds to buy land outright. The Land Act 1877 limited the size of blocks that could be purchased to 320 acres (130 hectares), and excluded anyone who already owned at least 640 acres (260 hectares). Most importantly, however, it allowed settlers to obtain land through deferred payment arrangements. The 1885 amendment to the Land Act also allowed settler associations to establish special settlements of smaller farms, based on deferred payment or perpetual lease arrangements.[5] The viability of smaller farms was further bolstered when, in the 1880s, the frozen-meat trade became established. Sheep meat, not just wool, became a valuable product that could be shipped to overseas markets – primarily to Great Britain. As a result of this development, land was farmed more intensively, and farming practices were fine-tuned, thereby improving productivity. The industry grew quickly: in 1895 New Zealand exported 2.3 million sheep carcasses, but by 1910 that number had more than doubled to 5.8 million.[6]

Large areas of the North Island remained in Māori ownership in the latter half of the 19th century. These lands were still forested and undeveloped, and, in official parlance, were referred to as 'waste lands'.[7] With the objective of transforming these great tracts of forested lands into productive farmland and settlements, Julius Vogel's government established an immigration scheme to bring in settlers from Europe willing to undertake bush-felling and public works.[8] This was the case on the eastern side of the Tararua and Ruahine ranges, too, where a large number of Scandinavian settlers were brought in to clear the Seventy Mile Bush, a large forested area stretching from Wairarapa to central Hawke's Bay.

Within the Manawatū region itself, it was the relatively open country of the coastal area around the mouth of the Manawatū River that was first settled by Europeans,[9] initially by whalers and traders, and subsequently by the New Zealand Company in an ill-fated land purchase in 1841 that eventually contributed to the company's downfall. The coastal Awahou Block, which centred around the town of Foxton, was later purchased by

This small portable mill was owned by brothers Jos and Bill Wilson, and operated between about 1898 and 1900. The brothers would cut out an area of forest, and then relocate the sawmill to another area. Here, the sawmill is pictured on the east bank of the Pohangina River, just south of Totara Reserve.

C.E. Wildbore, Palmerston North Library

the Crown in 1859. However, the block was relatively small and included none of the extensive area of forested land beyond the coast. It was to the latter that the government next turned its sights, purchasing this area in two large blocks, the Ahuatūranga (or Upper Manawatū) Block in 1864, and the Rangitikei–Manawatū Block in 1866. Together, these two blocks comprised about 450,000 acres, which extended from the Rangitikei River in the north to the foot of the Tararua and Ruahine ranges in the south, and to the Tasman Sea in the west.[10]

The Manchester Block, within the Ahuatūranga Block, was purchased by the Emigrant and Colonists' Aid Corporation in 1871, and the towns of Feilding, Halcombe and Ashhurst were developed and settled. But while Ashhurst was opened up for settlement in 1877 – with the intention of its

being a 'gateway' to the Pohangina Valley – the valley itself took much longer to develop, owing to its challenging terrain: rugged, densely forested hill country that was dissected by a river, which, like many New Zealand rivers, could swell to a raging torrent in minutes during heavy rainfall. On the one hand, roads and bridges were prerequisites to the valley's development, but on the other hand, they were costly and logistically difficult to build. For these reasons, the development of the valley did not begin in earnest until the 1890s. The clearance of bush by means of burning and sawmilling, the establishment of pasture and the building of roads and bridges took place, for the most part, during the 1890s and early 1900s.

Even sawmilling, the first phase of transforming bush country into productive farmlands and orderly settlements, relied on roads or other forms of access, such as tramlines.[11] For this reason, most of the sawmills in the Manawatū region were clustered around railway lines.[12] In 1879 an established sawmill-operator in the district, Peter Bartholemew, had set up a sawmill just outside of Ashhurst on the Pohangina Road, and, having leased 1200 acres of land from the Manchester Corporation, was cutting

Horses pulling a milled log across the Te Awaoteatua Stream bridge at the bottom of Church Hill, Awahou. Pictured are Mat Mai (front), who was Charles Wildbore's brother-in-law, and Trevor Madson (back), acting as brakeman. Not dated.

C.E. Wildbore, Palmerston North Library

mainly tōtara and mātai.[13] Once roading into the valley improved, sawmills sprang up wherever there was plenty of timber to cut and also a means of transporting it to a market.[14] In 1893 Adsett and Son set up a sawmill near Pohangina township,[15] and portable sawmills (similar to the one pictured on p. 15) were brought in for a period of time and then moved on to another area once the first area had been cut out.

After the most valuable and accessible timber was removed and milled, the settler farmers turned to 'breaking in' the land through a laborious process of bush-felling, burning and, later, stumping (the removal of tree stumps). On a large farm, this would be done in stages, and it could be several years before the whole of the farm was brought into production. While developing their farms, many settler farmers supplemented their meagre incomes through bush-felling on other landowners' properties, sawmilling, drain digging, working on railway construction and road making.[16]

Bush burning was a relatively inexpensive and readily accessible way of clearing land, and was therefore the most favoured method of clearance in the densely forested Manawatū region and throughout much of the North Island. However, bush burning also carried with it significant risks – both to the landowner and to the community. A sudden change of wind direction could cause the fire to spread to neighbouring properties, destroying fences and houses, killing livestock and even endangering the lives of the settlers themselves. In the mid-1880s, people became particularly sensitive to the risk of fires catching and spreading, following the disastrous conflagrations of 1885–86, which destroyed houses, businesses, farms and thousands of acres of bush in Taranaki and southern Hawke's Bay.[17] The settlers of the Pohangina Valley did not escape completely the calamitous consequences of a bush fire gone awry. In Umutoi, in January 1897, strong winds ignited the smouldering remains of a burn, and a number of settlers lost stock and fences to the flames – with some even losing their homes.[18]

The Pohangina Valley and Apiti were situated on the outer periphery of the 'frontier' of bush burning as the practice radiated out from the towns of the Manawatū region from the 1870s onwards. Bush burning did not become prevalent in these more remote areas until the late 1890s and the early years of the 20th century.[19]

Few questioned the necessity for bush burns, though some expressed concern about the scale and rapid pace of the destruction of forest. The Pohangina correspondent to the *Feilding Star* in 1890 observed:

> *We are looking out to be smothered in smoke for a spell as soon as the bush fires are started. There was one fire up the river the other day when a medium burn was made. However, if clearing goes on at its present pace, in a few years there will be no bush to burn, and the settlers will have to start growing blue gums for firewood and fences.*[20]

For a period, the burns gave rise to a thriving industry in the valley: grass seed production. Rather than graze stock, many settlers opted to grow and harvest grass seed for the first few years after the burn, with ryegrass,

Transporting grass seed by packhorse, circa 1904, taken on Thomas Harrison's property on Pohangina Road. Harrison is standing on the far right, Fred Hart is standing third from left (with hands on hips) and Jack Craven is driving the wagon.

C.E. Wildbore, Palmerston North Library

Men using hand sickles to reap grass, circa 1904. The man looking at the camera is Thomas Harrison, and the site is probably his farm on Pohangina Road, about two miles south of Utuwai School.

C.E. Wildbore, Palmerston North Library

crested dogstail and cocksfoot being the most common seed varieties. The seed was bagged up, and carried out by horse and cart to be sold to local seed merchants, including Barraud and Abraham of Palmerston North. So valuable was the crop that settlers arranged contracts with the local council to harvest seed from roadsides.[21] Initially, the grass yields were high, giving the impression that the land was highly fertile. However, this was an illusion created by the flush of nutrients released from the ash of incinerated forests. Once those nutrients had dissipated, the soil-fertility levels plummeted to more normal levels, and the native 'weeds', such as bracken fern and bidi bidi, re-invaded, along with introduced species, such as scotch thistle.

Through these processes of forest-milling, bush burning and the sowing of pasture, the Pohangina Valley was transformed, in a mere two decades, from dense, impenetrable forest to farmland and a string of small settlements.

THE ART OF THE BUSH BURN

A bush burn, circa 1904. The burn is on the property of Mr Roberts in Mount Richards, taken from the second terrace above the Pohangina River looking in a southwesterly direction. The fire burned between Coal Creek and Pohangina Road. This is the photograph for which Charles Wildbore is perhaps best known.

C.E. Wildbore, Palmerston North Library

Bush burning was often undertaken on a large scale and could be very risky, but it was far from indiscriminate. Before a burn was undertaken, the bush needed to be prepared through a labour-intensive process of bush-felling – and this was an art in itself. The first phase of bush-felling was 'underscrubbing'. This involved cutting down the dense understorey of the forest with axes or slashers. The understorey vegetation was left on the ground to dry over the summer months, later becoming fuel for the burn-off. The bush-feller would then cut the upper storey of the forest, using an axe or a saw, depending on the height, size and type of tree. An alternative technique was known as the 'drive', whereby several large trees standing in an approximate line were partly cut (scarfed) on one side, and then a driver tree was felled, setting in motion a line of falling trees.[22]

Settlers were advised to have the bush on their properties felled properly, as a good bush-felling was critical to a 'clean' bush burn. An optimal burn left little more than tree stumps and large limbs remaining – the rest of the bush would be incinerated, leaving a thick layer of ash. But if the bush did not burn cleanly, much of the vegetation would be left intact, requiring the farmer to either repeat the burn in subsequent years (thereby losing the opportunity to productively use the land in the meantime) or undertake 'logging up', a labour-intensive process of piling the unburnt logs and branches into large heaps for later burning.[23]

Bush-felling usually took place in the winter months so that the cut vegetation could be left over the summer months to dry out in readiness for the burn. The farmer then had to wait until the right weather conditions – usually in summer or early autumn – before 'firing up'. To be successful, a burn needed to be preceded by more than a month's dry weather, and, on the day of the burn itself, required a strong breeze in the direction desired for the burn.[24]

While there are few written accounts of bush burns in the Pohangina Valley, local newspapers carried vivid descriptions of burns in the Manawatū region. One such account in 1879 describes the spectacle created by bush burns in the countryside surrounding Palmerston North:

> *On Saturday, the country around the town seemed perfectly aglow with bush fires, while the atmosphere was most oppressive. The murky appearance of the sky, and the dull, muggy sensation experienced reminded one forcibly of the bush fires and hot winds of Australia, and as night closed in, the scene became grandly picturesque. On one side the glare of the flames lit up the heavens; while on another, the fire being only in an early stage, was to be seen sparkling here and there through the foliage of the trees, giving the appearance as if the bush hung*

Bush-felling at Komako, circa 1904. The men were employed by Dick Port (third from left, standing on jiggerboard with right hand on hip) and Alex Stewart, who had the contract to fell bush. Other individuals identified are Wilf Galer (left foreground, with arms folded), Bill Davey (third from right, in foreground, holding cross-cut saw) and Jack Enright (second from right, standing on jiggerboard in distance).

C.E. Wildbore, Palmerston North Library

Sowing grass seed, circa 1904. The handwritten description on the back of the photograph reads 'Mr Thos. Harrison, who had been so successful in this industry, is here shown sewing Kentucky Blue Grass'.

C.E. Wildbore, Palmerston North Library

> *with countless Chinese lanterns while again the bare and scorched limbs, twined in fantastic shapes, formed a weird foreground to the illuminated sky.*[25]

A successful burn enabled the farmer to sow seed immediately, and potentially to get stock onto the new pasture by the winter.[26] There was no need for him to clear logs or stumps from the land immediately because sheep and cattle were able to graze among them. Eventually, over a number of years, many of the logs decayed and rotted away, although farmers often undertook stumping to remove the larger, more obtrusive stumps.[27]

It was in this dynamic environment that Charles Edward Wildbore, a young man born to a working-class family in 19th-century London, found himself as the century drew to a close. How did he come to be in the Pohangina Valley, and what was it that compelled him to capture this landscape transformation as it happened? To answer these questions fully, we need to transport ourselves to Victorian-era England, three decades earlier, to explore Charles' origins.

The property of Charles Neilson on No. 1 Line, Awahou North, in 1896. His daughter, Eileen Savell, continued to farm the property until the 1960s, when she was in her 80s.

C.E. Wildbore, Palmerston North Library

Post office at Piripiri, circa 1897–1900. From left to right: Jack Enright, Mrs Culling, Mrs Fraser, daughter of the Culling family, Michael Enright (postmaster), Mr Culling and daughter. Other people are unidentified.

C.E. Wildbore, Palmerston North Library

Part 2

THE MAN BEHIND THE PHOTOGRAPHS

New beginnings

Charles Edward Wildbore was born in 1861 in Clerkenwell, an area in central London – the first-born child of Charles Lee and Emma Lydia Wildbore (née Guy). Charles Lee Wildbore himself was the eldest son of John Lee Wildbore, a former headmaster of a grammar school in Kirton in Holland, Lincolnshire.[28] The school fell into disrepute when it came under the scrutiny of a parliamentary inquiry and was found to be wanting, both in terms of its administration and its educational standards; John Lee Wildbore subsequently left for Canada in disgrace.[29] As a consequence, Charles Lee and his siblings did not benefit from the relative wealth and comforts that their father and grandfather had enjoyed, and, instead of gaining an academic education, as his father had, Charles Lee was apprenticed as a blacksmith in his teens, and went on to work as a farrier.[30] At some point, Charles Lee moved to London to find work as a farrier.

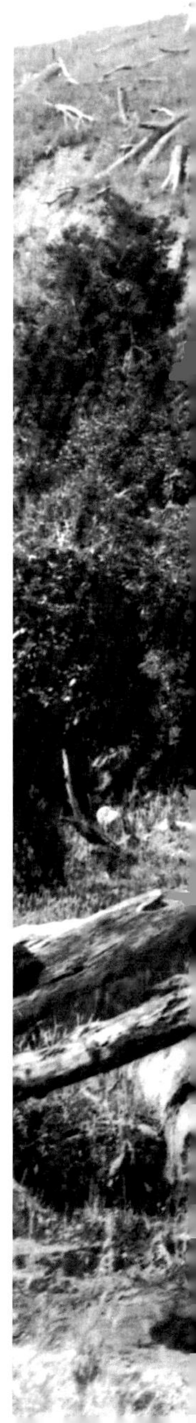

The Wildbore family at Beehive Creek circa 1897. Jane Wildbore is sitting in the dray in the background. Ruby is standing in the foreground, on the right. To her left may be her younger brother Percy. Behind them is likely to be Kathleen.

C.E. Wildbore. Lindsay Family collection.

Charles Lee had already been married once by the time he moved to London, where he met his second wife, Emma. His first wife had died less than a year after their marriage, suggesting that perhaps she died in childbirth. Emma, who was working as a laundress until she married and gave birth to Charles Edward, was the daughter of a watchmaker, Edward Guy.[31] Given her father's occupation, it is likely no coincidence that Emma's family lived in Clerkenwell, a district of London that was famous as a centre for watchmaking.[32]

Emma and Charles senior married in 1861 at All Saints Church in Islington, a central district of London, and took up lodgings on Trinity Street, south of the River Thames. Trinity Street was just one block from the county gaol, so it was unlikely to have been a desirable address at the time.[33] By the time Charles was born, also in 1861, Clerkenwell and the wider area of Islington comprised a working-class district of greater London, and was blighted by overcrowded and substandard housing conditions. However, its state of degeneration in the 19th century belied more illustrious origins.

The name 'Islington' had evolved from the old Saxon name, 'Gisla's hill',[34] which was bestowed upon the settlement in medieval times. At the foot of the hill, there were many springs, and this bountiful supply of fresh water made the area ideal for market gardens to supply the burgeoning populace of nearby London in the 17th and 18th centuries. Islington also became a popular destination for excursions by Londoners, who were attracted by its rural feel and opulent manor houses. The name 'Clerkenwell' also had a water-related provenance; the settlement was named for the Clerks' Well, a water source that adjoined St Mary's Nunnery in the 12th century. In the Middle Ages, Clerkenwell was favoured as a place to build monasteries and other institutions, such as religious schools, owing to its plentiful fresh water, open fields and proximity to London.[35] In the 17th century, Clerkenwell became a fashionable resort, with entertainments such as spas and theatres, as well as a 'musick' house.[36]

During the 19th century, Islington's population grew exponentially. At first, Islington was considered to be a fashionable district, attracting clerks, artisans and professionals, but by the middle of the century, many of the poor who had been living in central London were displaced by the development of new railway stations and goods yards, and they relocated to Islington. The area fell into a decline that lasted well into the middle of the 20th century.

Soon after Charles' birth, the family moved to Strood in Kent, about 40

Quinn's Buildings in Popham Road, the road in which Charles lived with his family in 1871, three years before they left England for New Zealand. The photograph was taken in 1967, but the tenement building was built in the 1870s and was typical of tenements in that area at the time.

Islington Local History Centre

CHARLES THE NOVELIST AND CHARLES THE BOY

One of the stories about Charles Edward Wildbore that entered family lore (presumably promulgated by Charles himself) was that, as a boy in London, he would tend to Charles Dickens' horse while the famous writer visited his favourite drinking establishment. Unfortunately, the story as we know it today is scant on detail – we do not know where this was supposed to have happened or under what circumstances. Nevertheless, although unlikely, it is not entirely inconceivable that Charles the boy and Charles the writer crossed paths.

The orbits of Dickens and Charles did indeed overlap, in both time and place. Dickens' life spanned the period February 1812 to June 1870, making Charles nine years old at the time of Dickens' death. Furthermore, Clerkenwell, where Charles was born and spent some of his childhood, was a place where the novelist spent

miles (64 kilometres) east of London. They lived there for a number of years before returning to Islington, possibly in 1869. The census in 1871, when Charles was nearly 10 years of age, finds them living in the Popham Estate, in Islington – a cramped and run-down tenement housing estate for the working class (see photograph on previous page).[37] Circumstances appear to have taken a turn for the worse for Charles senior and Emma – they had a lot of time throughout his life. Dickens' first home with his new family was in Doughty Street, Bloomsbury, only a short distance from Clerkenwell. In the early part of his career, Dickens frequently visited George Cruikshank, an illustrator, who lived in Amwell Street in Clerkenwell.[38]

The Clerkenwell area also provided the backdrop for scenes in a number of Dickens' novels and other stories, including *Oliver Twist*, which was originally published in monthly instalments between 1837 and 1839. Most notably, Clerkenwell Green was the setting for one of the pivotal events in *Oliver Twist*: the title character's initiation into the art of picking pockets. With characteristic irony and dry wit, Dickens describes this location as an 'open square in Clerkenwell, which is yet called, by some strange perversion of terms, "The Green"'.[39]

Oliver Twist flees the scene of a pickpocketing crime at Clerkenwell Green. An illustration by George Cruikshank in The Adventure of Oliver Twist.

Wikipedia Commons

Dickens' descriptions of Clerkenwell did not paint a favourable picture of the area, to say the least. Of Oliver Twist's initial encounter with Clerkenwell and adjacent areas, Dickens wrote:

> *A dirtier or more wretched place he had never seen. The street was very narrow and muddy, and the air was impregnated with filthy odours.*
>
> *There were a good many small shops; but the only stock in trade appeared to be heaps of children, who, even at that time of night, were crawling in and out at the doors, or screaming from the inside.*
>
> *The sole places that seemed to prosper, amid the general blight of the place, were the public-houses.*[40]

four children under the age of 10 years to support, and, according to the census, Charles senior was out of work. The couple had suffered the death of one child – Harriet Eliza – who was born in 1864 but died before she reached the age of one. Two years later, Emma Susan, who was one year old at the time of the 1871 census, would also die, aged three.

While the causes of the children's deaths are not known, what we do know is that London in the 1870s was far from being a healthy place – particularly if a person was poor. In 1866, when Charles was five years old, a cholera epidemic had swept through London – mainly affecting the poorer classes in London's East End. It was the fourth and the worst cholera epidemic in the city's history, claiming 5596 lives.[41] The 1870s were also a tumultuous time of change for both the economy and society as a whole: agriculture's golden age had passed, and many farm labourers and associated tradespeople were out of work. At the same time, industrialisation meant that machines were taking the jobs that people had been employed to do. Many people found themselves unemployed, and, with no social welfare in place to support them, were struggling to escape from the downward spiral of poverty in which they found themselves. It is not clear whether Charles senior was the victim of a transitioning economy or some other misfortune, such as sickness or injury – or whether he was simply between jobs at the time of the 1871 census.

Given the circumstances in which the Wildbores found themselves in the early 1870s, it is little wonder that Charles' parents were eager for a way to escape a future in which the spectre of poverty and hardship loomed large.

Montage of sketches depicting life on board an emigrant ship in 1875. The montage (clockwise from top left) shows immigrants embarking at the London docks, scrubbing the decks, watching a passing ship, queuing at the surgery, catching an albatross, dealing with heavy seas and (centre) whiling away the hours. Published in The Illustrated New Zealand Herald, 9 April 1875.

Alexander Turnbull Library, Wellington, ref: MNZ-0661-1/4-F

Journey to the South Pacific

Thousands of miles away, in the southern Pacific Ocean, the economic tribulations of the newly fledged British colony of New Zealand were to provide the opportunity for the Wildbores to escape their situation. By 1870 a slump in wool prices and a declining goldmining industry had tipped the vulnerable economy into a depression, and the New Zealand Wars had tarnished the colony's reputation overseas.[42] The colonial treasurer, Julius Vogel, had an ambitious plan to rejuvenate the New Zealand economy. The plan centred around developing the country's railways and roads, and facilitating large-scale immigration and settlement, all funded by overseas borrowing. Vogel's goal was not only to generate an economic boom, but also to create a new social order and to transplant 'British civilisation' to New Zealand, by settling new immigrants on land purchased or confiscated from Māori. This was a bold plan, and was facilitated by the Immigration and Public Works Act 1870.

As part of its immigration policy, from 1873, the New Zealand Government offered to pay full fares of eligible immigrants to travel from the United

Kingdom to New Zealand, and it despatched 177 government immigration agents to England, Scotland and Ireland to promote New Zealand as a land of new opportunity. Newspaper advertisements and posters called for married agricultural labourers and single female domestic servants, provided they were 'sober, industrious, of good moral character, of sound mind and in good health'.[43] The recruitment drive was successful, and, between 1871 and 1886, nearly 290,000 migrants flooded into the country – a number equating to the size of the Māori population as at 1871. However, only about 60 per cent of these immigrants stayed.[44] In 1874 New Zealand received its single biggest influx of immigrants – a net migration of 38,000 people – a number unsurpassed for more than a century.[45]

On 28 April 1874, the Wildbores were one of many families who took advantage of the offer of free passage to New Zealand, and they joined the

great wave of assisted migrants. With the death of Emma Susan the year before, Charles senior and Emma had three surviving children when they left England. Charles was 13 years old, John was eight and Harriet was six, while Charles senior and Emma were 38 and 36 years old respectively, when the family set sail from London on board *SS Euterpe*.[46] Four months later, on 30 August, and having traversed three oceans,[47] they arrived in Wellington. From there, they travelled on the *Stormbird* to Whanganui.[48]

The Wildbores did not settle in Whanganui, instead moving south to the Rangitikei District. Charles senior and Emma wasted no time in getting involved in community affairs, becoming founding members of the recently instituted lodge of the International Organisation of Good Templars in Bull Town (later renamed 'Bulls') in February 1875.[49] The International Organisation of Good Templars had its origins in the 1850s in the United

Rechabite Hall, Crofton, Marton, circa 1870s, taken by William James Harding (1826–99). The Independent Order of Rechabites was an organisation founded in England in 1835 as part of the wider British temperance movement to promote total abstinence from alcoholic beverages.

Alexander Turnbull Library, Wellington, ref.: 1/1-000327-G

States, and was established to promote total abstinence from drinking alcohol. It had a structure modelled on Freemasonry and used similar rituals and regalia, but, unlike the Freemasons, it admitted both men and women. The first lodges of 'Good Templarism' were established by newly arrived immigrants in 1872, and many more mushroomed throughout the country over the following years.[50] The lodge in Bulls, along with the lodge established in Crofton, near Marton, were the first in the Rangitikei and Manawatū districts.

By 1876 the Wildbores had settled in Marton, where Charles senior established a blacksmith shop at the back of the house.[51] In May 1876, Emma gave birth to their first New Zealand-born child, Willie, but he died just three days later. Following their move to Marton, Emma had became an officer of the lodge of the Good Templars in the town of Crofton, situated just a few miles out of Marton. This town was historic – if only fleetingly – in that it owed its existence entirely to the temperance movement. William Fox, the second premier of New Zealand (and long-time member of Parliament for Rangitikei), was an ardent supporter of the movement. In the early 1870s, Fox decided to subdivide a small portion of the land he owned in the Rangitikei District to establish a town reserved for supporters of the temperance movement. However, the town failed to become a success and, today, Crofton is just a scattering of houses, barely distinguishable from the rest of the rural Rangitikei landscape.

Both Emma and Charles senior remained active in the temperance movement for years to come, becoming members of the Feilding lodge of the Good Templars upon their move to the Manawatū District.[52]

Trials and tribulations: Fire, bankruptcy and hard labour

In 1877 the Wildbores had moved to Awahuri, a few kilometres southwest of Feilding, where Charles senior again set up a blacksmith shop. Their seventh child, Frederick Lee, was born in May 1877. But, just as the Wildbores were establishing themselves in their new home and community, they were struck by misfortune. Early one morning in April 1878, Charles' blacksmith shop, where he and one of their children were sleeping, caught fire. Father and child both escaped unscathed but 'the house was burned to the ground, nothing but a mass of charred and smoking sticks remaining'.[53]

The only thing Charles was able to salvage was a chest of clothing; all his tools, books and stores were incinerated in the fire. To make matters worse, he was uninsured.[54] Financially, this was a major setback: the following month, he declared himself bankrupt, and in 1879 Charles Lee Wildbore was listed as a resident who had failed to pay his rates to the Manawatu Highways Board.[55] He was plagued by money troubles – including more visits to the bankruptcy court – for years to come.[56]

Having lost everything, it is unlikely that Charles senior would have had the means to re-establish his blacksmithing business. It is fortunate, therefore, that in early 1880 he was given the opportunity to operate the farrier services offered as part of a new business set up by a Mr Robert Walker on Foxton Line (now Pioneer Highway), near Palmerston North.[57]

With this new venture, it appeared that, at the age of 47, Charles Lee Wildbore had picked himself up from his latest misfortune with resilience and dignity.

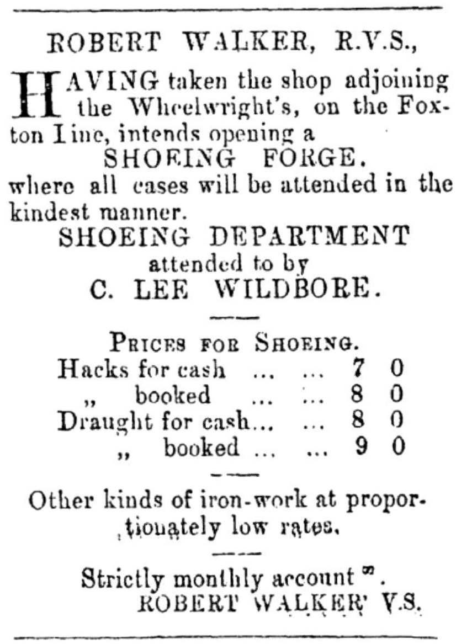

ROBERT WALKER, R.V.S.,

HAVING taken the shop adjoining the Wheelwright's, on the Foxton Line, intends opening a SHOEING FORGE, where all cases will be attended in the kindest manner.

SHOEING DEPARTMENT attended to by C. LEE WILDBORE.

PRICES FOR SHOEING.

Hacks for cash	7	0
" booked	8	0
Draught for cash... ...	8	0
" booked	9	0

Other kinds of iron-work at proportionately low rates.

Strictly monthly accounts.
ROBERT WALKER, V.S.

Advertisement for Mr Walker's shoeing forge, Manawatu Times, 17 January 1880.

Papers Past

In the mid-1880s, the Wildbores had moved to Taonui, to the east of Feilding, and Charles Lee (now in his 50s) and his son Charles had started down a new path – honey manufacture. This new enterprise was announced with great fanfare in the *Feilding Star* in October 1886:

> *To-day we publish a very important notice to Bee-keepers from Mr Wildbore of Taonui. Mr Wildbore has secured the best machine yet invented, for making the famous and useful comb foundations which have been applied to such immense advantage and profit in the United States of America. Bee-keepers on inspecting the samples will see at a glance the superiority of the article offered by Mr Wildbore.*[58]

Wildbore and son advertised themselves as 'dealers in bee-keepers' requisites', including hives and frames (with or without 'swarms') and honeycomb foundations. They also offered 'any information on bee culture'.[59]

At 17 years of age, Charles himself soon demonstrated abundant talent in producing honey, netting first prize at the Manchester Horticulture

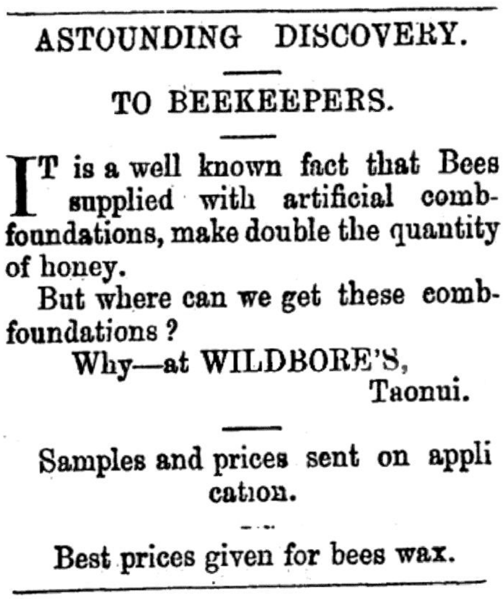

ASTOUNDING DISCOVERY.

TO BEEKEEPERS.

IT is a well known fact that Bees supplied with artificial comb-foundations, make double the quantity of honey.

But where can we get these comb-foundations ?

Why—at WILDBORE'S, Taonui.

Samples and prices sent on appli cation.

Best prices given for bees wax.

Advertisement for Wildbore's beekeeping supplies, Feilding Star, 19 October 1886.

Papers Past

Society for his honeycomb in 1888.[60] But just as the Wildbores' fortunes seemed to be on the rise, yet again they were visited by misfortune. In mid-September 1887, a cottage owned by Charles senior burned down, and all his equipment for the manufacture of honeycomb foundations, along with stores of beeswax, were destroyed. Once again, he was uninsured.[61]

And this time, events were to take a more sinister turn. Charles senior believed the fire was foul play, and he was convinced he knew who the culprit was.[62] It appears he had been on bad terms with a neighbour, Jeremiah Corkery, for some time. He believed that Corkery had lit the fire as retribution for Charles senior's having shot some of Corkery's pigs after they had wandered repeatedly onto his property (and it must be said, if this had been the case, Charles had given fair warning – he had taken out a series of notices in the *Feilding Star* warning that he would impound any horses and cattle, and shoot any pigs or dogs, straying onto his property).[63] On the evening of 8 October, a few weeks following the fire, Charles paid Corkery a visit. Their encounter soon escalated into a heated argument, culminating in Charles attacking Corkery with a knife. Corkery suffered a cut to the face, one of his fingers was cut off and his hand was badly lacerated.[64]

Later that month, Charles senior was found guilty of assault and was sentenced to nine months' hard labour.[65] He was 57 years old and had been suffering ill health, so this sentence would have seemed like a lifetime. For his wife, who had a one-and-a-half-year-old son (William Alfred had been born in April 1886) as well as a six-year-old daughter (Louisa, born in January 1881) to care for, this would have been a terrible blow.

If this incident had a silver lining, it was the upwelling of support for Charles senior from the wider community following his sentencing. It is evident that sympathies were with him, despite his having been found guilty. Soon after he was sentenced, a petition was initiated, 'praying for a mitigation of the sentence', which many in the community felt was too harsh. By November 1887, several hundreds of well-wishers to the family had signed the petition.[66] But, unfortunately for the Wildbores, their supporters' efforts were in vain – in March the following year, the inspector of prisons announced that he would be declining the request.[67]

Nor was the 1887 fire the last to inflict tragedy on the lives of the

Wildbores. Nearly 20 years later, fire was to lead to the death of Emma Wildbore. In November 1916, at the age of 79, she was discovered unconscious on the path outside her house, with all her clothes burned from her body. It appeared her clothing had caught fire from the oil stove in the kitchen.[68] It is perhaps fortunate that her husband was not alive to endure this tragedy; he had died of natural causes just months before, in June 1916.

Bailey's Sawmill, near Taonui, 1890. This was one of a number of sawmills operating in the area at the time.

Manawatu District Libraries

Charles Edward Wildbore strikes out on his own

Meanwhile, against this at times tumultuous backdrop, young Charles Wildbore was steadily striking out towards social and economic independence. In his late teens and early twenties he spent several years working at sawmills in Taonui.[69] In other ways, too, he demonstrated initiative and resourcefulness – albeit with a perilously casual regard for rules at times (see inset, next page) – in his eagerness to make a contribution to family life.

WILDBORE VERSUS WILD DUCK

In 1879, at the age of 17, Charles had a brush with the law in an incident that reveals an endearing mix of audaciousness and honesty. He was caught hunting wild ducks on horseback before the season had been declared open, in contravention of the game laws, and, in a stroke of bad luck, his transgression was witnessed by none other than the local secretary of the acclimatisation society (the body responsible for policing the game laws). Conscious of the widespread lack of respect for these hunting restrictions, the society decided to press charges against Charles to make an example of him.

> WILDBORE VERSUS WILD DUCK.—We understand that Constable Gillespie will perfer a charge of Breach of the Game Laws against a young man named Wildbore who in defiance of the close of the season has been making it hot for the wild ducks in the neighborehood. The case has been brought at the instance of the Acclimatisation Society, and will be heard at the Resident Magistrate's Court to-morrow.

'Wildbore versus Wild Duck', a news item in the Manawatu Times, 12 March 1879.

Papers Past

Somewhat at odds with its stern intent, the acclimatisation society requested that the court not inflict a heavy penalty (an offender was liable for a fine of up to £20 for such an infringement) on account of Charles' complete honesty about his actions and the fact that he had no game in his possession when he was caught (suggesting that he was perhaps not a very good shot!). In the end, Charles got off lightly – he was fined five shillings as penalty, and nine shillings for costs.[70]

In the mid-1880s, when he was in his twenties, Charles became a member (and officer) of the Feilding branch of the Independent Order of Odd Fellows, suggesting altruistic, community-minded inclinations.[71] The incongruously named organisation, otherwise known as a 'friendly society', was a precursor to state social welfare. Friendly societies had their roots in the English craft guilds of the 13th century, and were established by socially conscious individuals during the Industrial Revolution to provide working men and their families with financial support. New Zealand's first friendly-society lodge was established (by the Order of Odd Fellows)

in 1842 in Nelson, and soon lodges had been set up in almost all the main centres. By 1879 the order had 113 lodges and over 8000 members, although membership declined after state welfare was introduced in the 1930s.[72]

In November 1888, at the age of 27, Charles married Jane Emily Dallison (also known as Emily Jane).[73] Jane was a local girl – born in Bulls in 1869 to an English father and a New Zealand-born mother.[74] Her father, Thomas Rought Dallison, worked in sawmilling and was later employed as an engine driver at Taonui.[75] Jane was 19 years old when she married Charles, eight years his junior. It is possible that her having fallen pregnant to Charles may have expedited their marriage plans – she was about four months pregnant at the time of their nuptials (the baby, Charles Henry, was born in April 1889). The couple went on to have 13 children, though their last-born, Thomas Norman, died at the age of three.[76]

The Wildbore family, circa 1915. From left to right: Evangeline (Eva), Sylvia, Percy, Baden, Stanley, Charles (seated), Cecil, Charles, Reginald, Owen, Jane (seated), Mana, Kathleen and Ruby.

C.E. Wildbore, Palmerston North Library

During the 1920s and 1930s, Charles and Jane also looked after three of their grandchildren while they attended school in Pohangina Village. In 1922 Ivan Clarence Lepper came to live with them after his mother, Kathleen (second from right in the photograph opposite), died in 1922 at just 30 years old.[77] In the late 1920 and early 1930s, Mervyn Williams (Evangeline's son) and Beverley Wildbore (Reginald's daughter) also lived with their grandparents. In Mervyn's case, this was the result of his parents, Roy and Evangeline Williams, experiencing financial difficulties during the Great Depression. Tragically, in the case of Beverley, it was the consequence of her mother, Ida, dying in 1934, aged only 25 years old (Beverley was five years old). She also lost her father, Reginald, only seven years later, when the POW ship he was being transported on was torpedoed by Allied forces.[78]

By all accounts, the couple's warmth and generosity extended beyond their own children and grandchildren. They appear to have had an 'open door policy' for all the children of the village; their children's friends were always welcome to share food in the kitchen, for example. Even after leaving the family home, the couple's children and grandchildren visited frequently,[79] so it can be imagined that the Wildbore home was a perpetually lively household, in which the aromas of baking, in anticipation of innumerable hungry mouths, were always evident.

Following his marriage and the birth of the couple's first child, Charles no doubt turned his mind to where the family would settle and how he would make his livelihood. The following pages set out how the family came to live in the Pohangina Valley, and how Charles' photographic activity became interwoven with the dramatic transformation of the valley.

Part 3

CAPTURING A PLACE AND TIME

A pioneering settler of Manawatū's hinterlands

The Pohangina Valley, about 15 to 20 kilometres from the Wildbore family home in Taonui as the crow flies, is named for the river that has carved its course through the land over millennia. (Refer to the map on page 82.) The Pohangina River flows south from its source on the western flanks of the Ruahine Range and converges with the Manawatū River near Ashhurst, where that river emerges from its deeply incised gorge through the Ruahine and Tararua ranges. The Pohangina Valley formed much of the northwestern corner of the Ahuatūranga Block, an oblong-shaped tract of land stretching from the present-day settlement of Linton in the south, to Apiti in the north, with

The settlement of Komako, Pohangina Valley, looking south down the river, 1907. The house on the far left was originally built for employees of Bartholemew's Sawmill in the 1890s, before becoming the blacksmith's residence. The white house (centre) on the right-hand side of the road was the Phelans' home, and was originally the Rimmer and Craven Store in the 1890s. The small building in the foreground was a hut where the teamsters who drove for the Sash and Door Company stayed overnight, and the long building (lower right) was used as a stable for their horses.

C.E. Wildbore, Palmerston North Library

the upper Manawatū and Pohangina rivers flowing south along its 'spine'. When government surveyor J.T. Stewart undertook his initial survey of the Ahuatūranga Block in 1859, he described it as 'mostly bush land' with only a few natural clearings – one of which was located at the future site of the present-day town of Ashhurst.[80] The New Zealand Government purchased the block in 1864 from tangata whenua: Rangitāne, Ngāti Kauwhata and Ngāti Tumokai tribes.[81]

Subsequently, at the western entrance to the Pohangina Valley, the settlement of Ashhurst[82] was established in 1877, as part of the Manchester Block. This was a special settlement scheme promoted by the Emigrant and Colonists' Aid Corporation, and headed by the Duke of Manchester.[83] Beyond the boundary of the Manchester Block, a large portion (about 25,000 acres) of the area around the future settlement of Pohangina was vested to the Wanganui Harbour Board in 1877.[84] The purpose of the endowment was to enable the harbour board to use the proceeds from the subdivision and sale of the land to finance improvements to Wanganui Harbour, for which it did not have adequate funds.[85]

Plan of the Wanganui Harbour endowment, 1881, surveyed by A. Dundas.

Alexander Turnbull Library, Wellington, ref. MapColl 832.43gbbd 1882 9470

However, the Wanganui Harbour Board found itself in a 'chicken-and-egg' situation, as it proved to be very difficult to attract prospective purchasers when they would have no means by which to access their land.[86] By 1879 there was 'a good metalled road' between Palmerston and Ashhurst, and another was being completed between Feilding and Ashhurst, but there were no roads beyond Ashhurst into the Pohangina Valley.[87] The Feilding Corporation had plans to complete a road from Ashhurst towards Pohangina, but only to the boundary of its block at Raumai. With that in mind, the corporation urged the Wanganui Harbour Board to continue the road into their endowment block, emphasising 'the importance of the main line of roadway as far as Ashurst [*sic*] being carried out with all speed since it would bring the totara flats in the Pohangina Valley within reach of a market'.[88] Despite these urgings, there were long delays in even getting the road lines surveyed, let alone constructed.

Meanwhile, eager to make some earnings on the land vested to it, the harbour board pressed ahead with its efforts to put sections up for sale. In January 1882, sections in Block IX (just west of the future site of Pohangina Village) and Block XIII (south of Valley Road) were first offered for sale, for cash or deferred payment, under the provisions of the Land Act 1887 (see 'Plan of the Wanganui Harbour endowment' opposite page). But only two

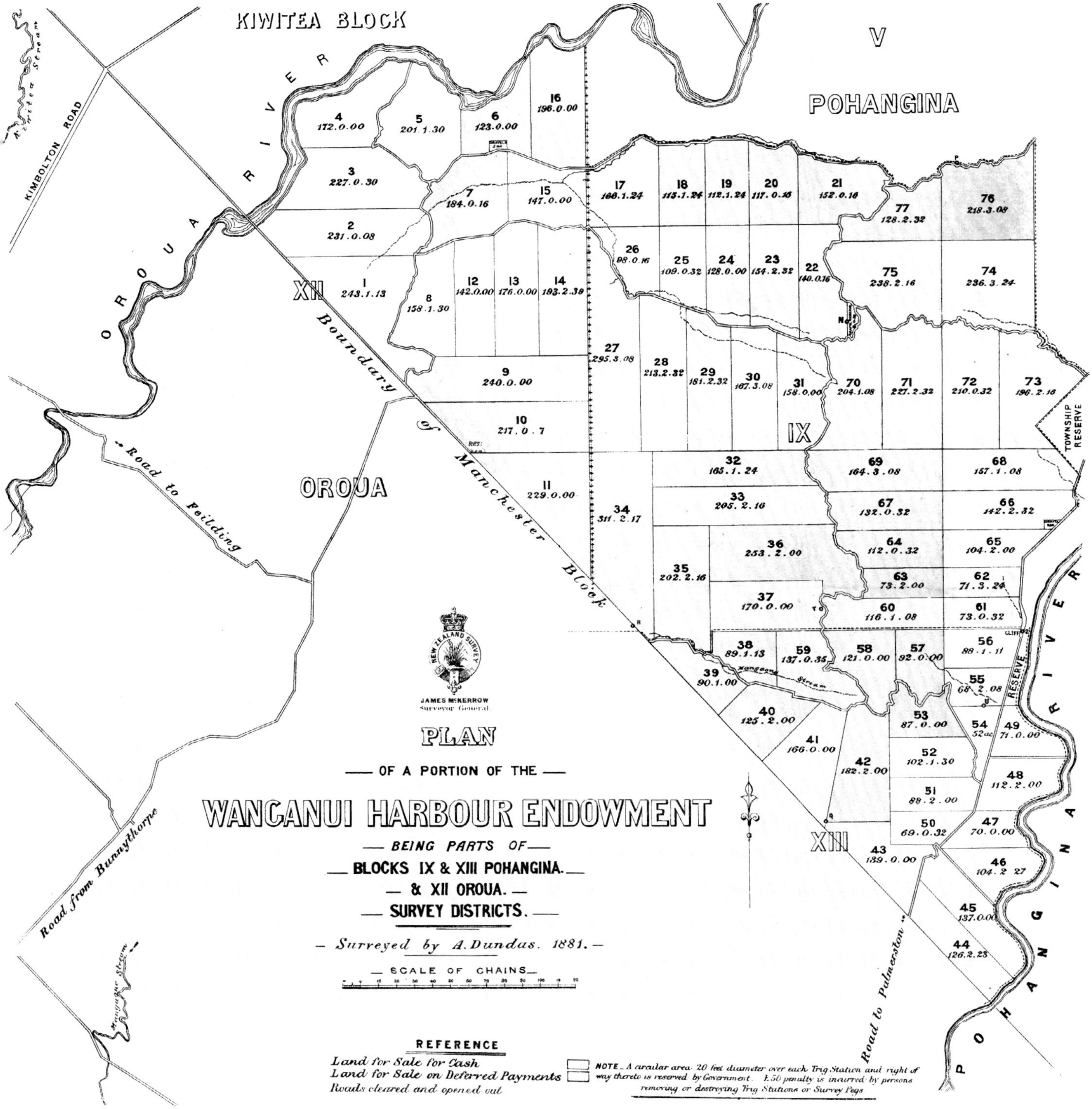
KIWITEA BLOCK
V
POHANGINA
KIMBOLTON ROAD
OROUA RIVER
XII
Boundary of Manchester Block
IX
XIII
OROUA
Road to Feilding
Road from Bunnythorpe
Road to Palmerston
POHANGINA RIVER
TOWNSHIP RESERVE
RESERVE
NEW ZEALAND SURVEY
JAMES McKERROW
Surveyor General.
PLAN
OF A PORTION OF THE
WANGANUI HARBOUR ENDOWMENT
BEING PARTS OF
BLOCKS IX & XIII POHANGINA.
& XII OROUA.
SURVEY DISTRICTS.
Surveyed by A. Dundas. 1881.
SCALE OF CHAINS
REFERENCE
Land for Sale for Cash
Land for Sale on Deferred Payments
Roads cleared and opened out
NOTE _ A circular area 20 feet diameter over each Trig Station and right of way thereto is reserved by Government. £.50 penalty is incurred by persons removing or destroying Trig Stations or Survey Pegs

sections in Block XIII sold. Land in both blocks was again offered for sale in March and December of that year, with about seven sections sold in the latter sale.[89]

The slow sales would not have been helped by the lack of progress made on roading in the 1880s.[90] The failure to build roads into the Pohangina Valley was not only impinging on sales of the harbour board's own land; it was also preventing access to government land beyond the perimeters of the harbour board's block.[91] There were glimmers of progress when the Kiwitea Road Board, formed in about 1882, took over responsibility for road making in the Pohangina Valley. It made some headway on extending existing roads and constructing new ones into the endowment block over the following years.[92] A clear sign of progress, too, was the opening of Pohangina School in the village in 1887, initially with an average attendance of 24 students daily.[93] By 1890 the Wanganui Harbour Board was apparently eager to divest itself of the burden of the remaining sections, as they were advertised for sale at reduced prices.[94]

Plan of suburban and town sections, township of Pohangina, 1891. Surveyed by F. Gillett, 1891.

Alexander Turnbull Library, Wellington, ref. 832.43bje 1930

In 1891 the Pohangina Small Farm Association was formed,[95] and, soon after, Charles Wildbore became a member,[96] apparently with the intention of purchasing land in the Pohangina District. As noted previously, small farm associations were created under the Land Act 1885, and allowed settler associations to establish special settlements of smaller farms, obtained on deferred payment arrangements or perpetual leases.[97]

In March 1893, Charles won the ballot for a 150-acre section in the newly opened Pohangina Block IV,[98] situated on the eastern side of the Pohangina River, just north of Komako (in the Umutoi Riding) (see 'Provisional plan of Pohangina County', p. 83).[99] His brother John also won the ballot for a 150-acre section in this block, but later forfeited it.[100]

It may have been that farming on this scale did not suit Charles: in early 1896, he sold his farm block in Umutoi and bought a 'surburban' section (of 13 acres) in Pohangina township (section 2 on the 'Plan of suburban and town sections, township of Pohangina', opposite).[101] In 1897 he came to a lease arrangement with the Wanganui Harbour Board for Block I, a five-acre parcel of land that stood between his existing block and Pohangina Road. This arrangement evidently suited the harbour board because even though it planned to subdivide the land into 20 quarter-acre sections (as shown in the 'Plan of the Wanganui Harbour endowment', on p. 45), it did not intend to do this for another several years.[102] Later, Charles also leased

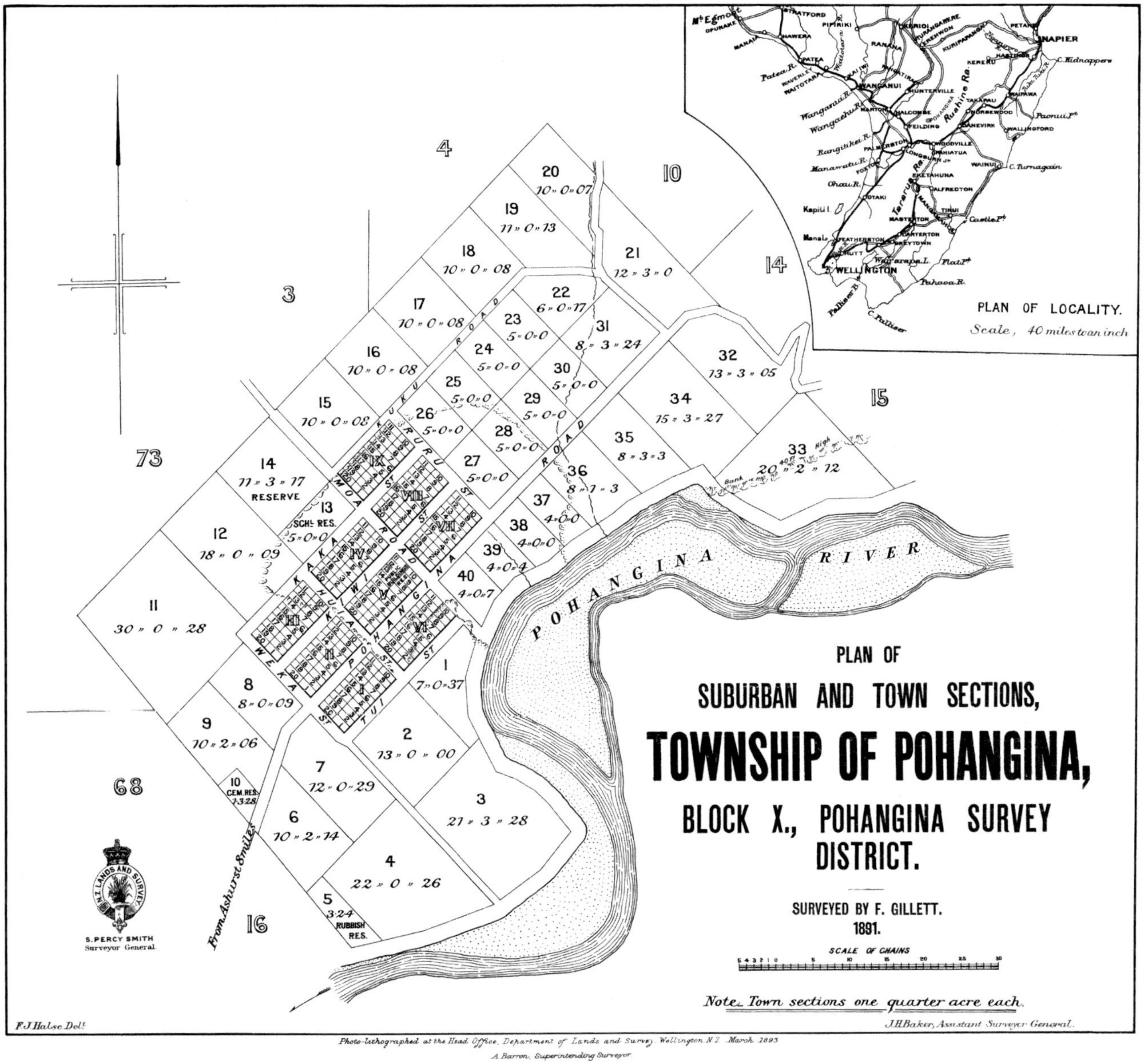

the cemetery reserve land until it was vested back to the Pohangina County Council in 1914.[103] Further afield, he leased land from the Wellington Land Board in Nireaha Village, near Eketahuna, in Wairarapa.[104]

Around 1902, Charles sold the 13-acre section and bought a number of

The 'main' street of Pohangina Village, Finnis Road, in autumn colour. The road sign at far right points to Kaka Street, and the street sign to its left and further in the distance points to Kiwi Street. The Wildbore home was situated between these two streets. *Catherine Knight*

town sections, including eight quarter-acre sections in the block between Kiwi and Kaka streets.[105] It was on Kiwi Street that the family settled and built their family home. The balance of the sections on which he paid rates were on the south side of Pohangina Road; these were located where the Pohangina Wetlands are today (see p. 73), and near where the Methodist church now stands (see opposite page). It is likely that these are the sections he leased from the Wanganui Harbour Board.[106]

By 1918 Charles' accumulation of land in and around Pohangina Village appears to have reached its maximum, with about 29 town and suburban sections listed against his or his wife's name in the rate book.[107] The

Methodist church, Pohangina Road. The church was opened in 1907. *Catherine Knight*

suburban sections were larger blocks on the periphery of the village and were likely used to graze his stock (the Wildbores farmed a small herd of dairy cows) or perhaps to place beehives. However, it is not clear what Charles did with the town sections, particularly as a number were not contiguous. It could be that they were purely an investment to on-sell for a profit when the time was right, though his retention of much of the land until his death in 1937 suggests that either this was not the case or the opportunity never eventuated.[108]

Charles took an active role in community life and organisations, becoming a committee member of the Pohangina Farm Homestead Association and the Pohangina School committee, and serving on community boards such as the Pohangina Domain Board.[109]

An advertising poster designed and photographed by Charles. The house in the photograph is the Wildbore home.

C.E. Wildbore, Palmerston North Library

One of the Wildbore children feeding the chickens in the yard. Some of Charles' beehives can be seen in the background, and a large swarm of bees can also be seen in the tree in the foreground.

C.E. Wildbore, Palmerston North Library

Continuing with the tradition established by his partnership with his father, Charles also established a honey business – the Pohangina Valley Apiary. He made his own hives and produced his own marketing material for the apiary. He sold his honey in two- and four-pound tins and in kerosene tins to retailers, such as Taylor's General Store in Hunterville.[110] His honey continued to win prizes, including at the Feilding Exhibition in 1903 for 'best honey in comb'.[111] One anecdote recounted about Charles is that, one afternoon, he found a swarm of bees on a roadside near the forest reserve (now Totara Reserve). Without hesitation, he took off his shirt, dropped the bees down the sleeve and took them home. Pohangina residents also remember being captivated as children when he showed them the queen bees he was sent by other apiarists, which he stored in matchboxes in which he had placed a bed of cotton wool.[112] He was obviously recognised as an enduring expert in this field because, in 1927, when he was in his 60s, he was invited to speak to the Dairy Farmers' Union at Pohangina on the subject of 'practical bee-keeping for farmers'.[113]

Like many men of this era, Charles also dabbled in inventing things – an interest he pursued over at least two decades. He invented a folding dish cover, and was granted a patent for this invention in 1895.[114] Together with a J. McLaurin, he also developed a technique for testing bales of wool, for which he applied for a patent in 1906.[115] He also developed a fire alarm, applying for a patent for this invention in 1913.[116] (It is difficult to discount the possibility that Charles' motivation for inventing an effective fire alarm may have stemmed from his parents' unfortunate experience with fire.)

Charles worked as a skilled labourer, relying mainly on contract work to earn an income, while Jane was primarily responsible for farm duties. One of the more significant contracts he undertook, along with another Pohangina man, involved moving and re-erecting the harbourmaster's house at Foxton Harbour.[117] The location was 80 kilometres away by road, so it is likely that he stayed on-site for the duration of the contract.[118]

Wildbore discovers the wonders of photography

It is in the early 1890s that we see the first signs of Charles' interest in photography. This emerging interest coincided with an era when photography was becoming more accessible to a broader section of the community – until that time it had been limited largely to dedicated professionals. Before the introduction of dry-plate photography in the 1880s,[119] photographs were made using the wet-plate collodion process, which required the photographic plate to be coated, sensitised, exposed and developed within a span of about 15 minutes.[120] This meant that a photographer who wanted to take photographs outside the studio had to take a portable darkroom with them so that the images could be developed on site, making this a technically challenging, laborious and expensive undertaking.[121] The dry-plate process involved glass plates pre-coated with a gelatin emulsion of silver bromide.[122] This process was not only more convenient, but the emulsion could also be much more sensitive, thereby greatly reducing exposure times.

The dry-plate process was in turn superseded by the introduction of celluloid film. The use of roll film was pioneered by the Kodak film camera, which became available in New Zealand from the 1890s, and grew increasingly popular from the early 20th century.[123] Charles began using a roll-film camera for some of his work from around 1915,[124] but also continued to use glass-plate technology for some time. This is most likely

An unidentified photographer taking a photograph of a road-grading team, circa 1910. This photograph, one of a pair of stereoscopic images, shows the photographer using a focussing cloth and the photographic equipment typical of this time.

C.E. Wildbore, Palmerston North Library

because, in spite of the greater convenience of hand-held film cameras, he valued the quality that could be achieved using large-format, glass-plate cameras.

It is evident from what we know of his early photographic activities in the 1890s that Charles took a keen interest in community life. Photography was a way in which he was able to contribute to the community, by recording special events through his photographs.[125] Few of these early photographs survive, although the 'Pohangina Notes' section of the *Feilding Star* has accounts of his photographic activities at the Pohangina and Komako school picnics, as well as of his entertaining the children with magic lantern shows (an early form of image projector, using painted images or photographs on glass plates).[126] He also took photographs of local events, such as weddings.[127]

Much of Charles' photography appears to have been carefully planned (particularly his later landscape photography), but he was also opportunistic. If anything unusual or unexpected occurred, Charles would

The original Pohangina school at Holley's Hill (the first hill after the Raumai Junction when travelling north), circa 1894. The headmaster was A.E. Walsh. In 1895, the school was re-erected on Finnis Road in the Pohangina Village, where the building still stands (the school closed in 1972).

C.E. Wildbore,
Te Manawa Museum

The Scott's farm on No. 2 Line, circa 1900–10.

C.E. Wildbore, Te Manawa Museum

enthusiastically embrace the opportunity to capture the moment with his camera. In 1907, for instance, a rare Japanese lily, grown with great care by Pohangina Village resident Mrs Hart, suddenly burst into full bloom for the first time in nine years. The Pohangina correspondent to the *Manawatu Standard* reported that 'Mr Wildbore has secured some photographic views of this singular flower', which was described as standing 'nearly eight feet high, and [having] about forty large well-shaped blooms, white streaked with crimson'.[128] Sadly, his photographs of this 'singular flower' do not appear to have survived, at least not in any public collection.

From the early 1900s, Charles developed an interest in photographing landscapes, and it is from this time that we have a more tangible photographic record of his work. His talent obviously did not go unnoticed, because in 1903 Charles was commissioned by Barraud and Abrahams, seed merchants in Palmerston North, to take a series of photographs showing the processes of bush clearing, grass sowing and harvesting grass seed that

were transforming the forested hinterland of the Pohangina Valley into farmland. Charles produced this series of photographs around 1904, and it features the image for which he is probably most known – a great plume of smoke billowing from a bush burn on a farm at Mount Richards (see p. 20). This photograph subsequently won first prize at an international exhibition in San Francisco.[129]

From this time, too, he began entering his photographs in the annual Manawatu A&P Show photography competitions.[130] At the 1906 show, he won first prize for a photograph entered in the 'landscape studies' category (open to amateur photographers only),[131] and one of his photographs received a 'commended' at the 1907 show.[132]

It is likely that Charles submitted some of his photographs to newspapers for publication, possibly for payment. An intriguing example of this is

Country cottage. The photograph's date is unknown but the cottage is 1880s/1890s in style. The woman (at right) wears clothing circa 1890s–1905 in style.

C.E. Wildbore,
Te Manawa Museum

Two versions of 'The travelling tinker', circa 1903 (it appeared in the Otago Witness in 1905). The 'tinker' is Charles Wildbore, with three of his children (from left to right), Kathleen, Mana and Ruby, looking on. The only difference between the two images is the volume and shape of the smoke plumes rising from the 'fire'. In what appears to be the original image, the metal drum to the right of the frame simply contains fragments of wood, perhaps in readiness for a fire, but with no fire burning (see page 80).

C.E. Wildbore, Museum of New Zealand Te Papa Tongarewa (above) and Palmerston North Library (right)

his photograph entitled, 'A travelling tinker and his interested onlookers', which appeared in the *Otago Witness* in July 1905. It turns out that the photograph was entirely staged: the 'travelling tinker' was Charles himself mending a pot, while the 'interested onlookers' were his children, Kathleen, Mana and Sylvia (along with the Wildbore cat).[133]

What makes this photograph even more fascinating is that three versions of it are known to exist, with the point of difference being the volume and shape of the smoke rising from the 'fire'. In what appears to be the original image, the metal drum to the right of the frame simply contains fragments of wood, perhaps in readiness for a fire, but with no fire burning.[134] The two other variations have varying amounts of smoke emanating from the drum, an effect he created by placing cotton wool on the glass plate before developing the print.[135] Charles apparently took the photograph using a 'trip-wire' (a long cable release for the shutter) around his foot. The wire ran from his foot around the back of the drum to the camera.[136]

It is clear that not everyone in Charles' family celebrated his interest in photography. His grandson recalls how Jane was palpably frustrated by her husband's preoccupation with photography, and did her best to 'dissuade' him from pursuing it. She would bury the glass plates, or have the grandchildren scrape them clean. And it seems that being the subject of his photographs was not universally welcomed by his children either. Later in life, his youngest daughter, Evangeline (Eva), would do her utmost to avoid being photographed, apparently weary of being photographed as a child.[137]

What is beyond debate is that Charles had a talent for photography. He had a well-attuned eye for composition, and, judging from his photographs, spent a lot of time carefully arranging the elements of each frame, including its human subjects. As Athol McCredie, curator of photography at Te Papa National Museum,[138] observes of Charles' skill in composition:

> *All of this involved substantial time and effort to place things, animals and people into harmonious arrangements. None of these would have just been sitting there for a casual snapshooter to capture. The images all had to be planned out and people organised into photographic compositions. This is what distinguishes a good photographer.*[139]

We have no information as to how and where Charles obtained his

Post office at Piripiri, circa 1897–1900. From left to right: Jack Enright, Mrs Culling, Mrs Fraser, daughter of the Culling family, Michael Enright (postmaster), Mr Culling and daughter. Other people are unidentified.

C.E. Wildbore, Palmerston North Library

cameras and photography equipment, nor how he gained his knowledge of photography. It is possible he was given a camera, or, having developed an interest in photography, he may have sourced his cameras and associated equipment locally.[140] By the late 1880s a number of shops in the Manawatū region and also further afield, such as in Woodville or Whanganui, were selling cameras and photography equipment. In Feilding, for instance, 'Watchmaker, Jeweller & Optician' G.W. Fowles was advertising himself as a 'Dealer in Photographic Lenses, Cameras, Chemicals, dry plates, &c.' from the early 1890s,[141] while E.A. Haggen Bookseller & Stationer, Woodville, was an agent for The Eastman Dry Plate and Film Company, and was selling Kodak cameras from 1890. The pursuit of photography in the region received a boost when, in 1899 or 1900, well-known photographer Enos S. Pegler moved from Auckland to Palmerston North, where he established

a photographic supply store and also offered classes and guidance on photography.[142] Pegler remained in Palmerston North for seven years, before returning to Auckland in 1907.[143] However, by this time, Charles had established himself as a competent photographer, and it is unlikely he would have sought advice from Pegler, though he perhaps purchased equipment from him.[144]

It is also unclear what type of camera Charles used (unfortunately, none of his equipment appears to have remained in the family), but it is said that every time he had any spare money, he would spend it on a new camera (much to the chagrin of his wife Jane).[145] His first camera was likely to have used full-sized plates (6 ½ inches by 8 ½ inches or approximately 19.5 cm by 21.5 cm).[146] This format was standard in the 19th century, but its use gradually declined from the beginning of the 20th century. By the 1920s, it was mainly used for studio photography. This was because a whole-plate camera was larger and needed to be used with a tripod.[147] It was therefore much less convenient to carry and use than the hand-held cameras that became available from the beginning of 20th century. Charles developed all

Rural mail delivery circa 1907–09. The two-horse brake (carriage) on the right delivered mail between Raumai and Delaware Flat. Michael Enright (fourth from left) was awarded the initial contract, and his son Dan (holding the reins) drove the brake. Seated to Dan Enright's right is Jack Enright. A post office was established at Piripiri and, later, was moved across the river to Delaware Flat. This photograph was taken on Delaware Flat, south of Takapari Junction.

C.E. Wildbore, Palmerston North Library

Stereoscopic photograph of lakelet in the Square, Palmerston North. The lakelet was constructed in 1909, and the photograph is dated circa 1913. C.E. Wildbore, Te Manawa Museum

his photographs using chemicals he mixed himself in a dedicated shed (one of several sheds on his property), which served as a darkroom. It is said that he rinsed his photographs in the stream at the end of the street.[148]

It is also evident from Charles' photographic legacy that he acquired a stereoscopic camera sometime around or before 1910 – the two earliest stereoscopic photographs taken by Charles have been dated circa 1910.[149] Stereoscopic cameras took slightly different photographs of the same subject (usually a landscape or a scene, rather than a portrait of people), and when the printed photographs were viewed through a special stereoscopic viewer, they had the effect of appearing in three dimensions. In England, by the 1860s, no middle-class home was complete without a stereoscope and a selection of stereoscopic cards.[150] The craze did not

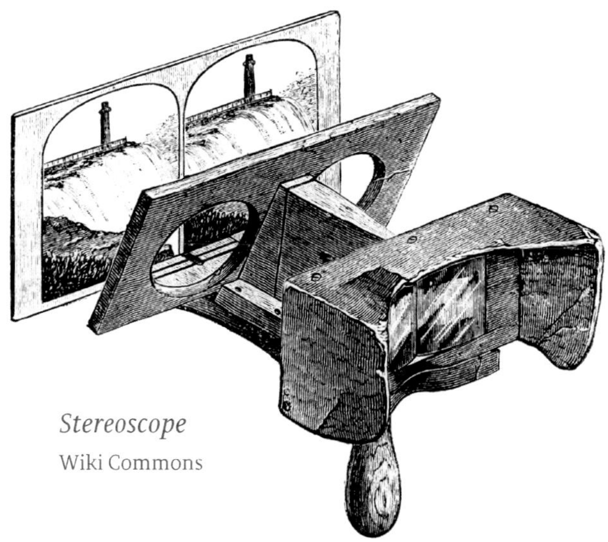

Stereoscope
Wiki Commons

THE MANGAHAO POWER SCHEME

take long to reach New Zealand. 'Stereoscopic views' were being sold in Nelson, Dunedin and Auckland as early as 1862.[151] The stereoscopic camera, on the other hand, seems to have taken some time to reach New Zealand's shores. There are some instances of private sales in the late 1890s,[152] but the stereoscopic camera only became available through a sprinkling of retailers

The fruition of the Mangahao hydroelectric power scheme was a significant event in New Zealand's history. It was the first government scheme built in the North Island, and it satiated (at least for a time) an urgent demand for electricity in the lower North Island. Support for the scheme was particularly energetic in the Manawatū region, where the local member of Parliament, W.H. Field, pressed for the scheme's construction.[155]

Construction of the scheme began in 1920, and was completed four years later, in 1924. The No.1 Upper Mangahao dam was subsequently constructed between 1925 and 1927. The Hutt River had originally been mooted to supply electricity for the Wellington area, but the Mangahao River, with its headwaters in the Tararua Range, was eventually chosen because it was considered to have more potential, and also it was better placed to supply power further north, to as far as Whanganui.[156] The scheme was an engineering triumph, because it diverted a river from its natural eastward course down the Tararua Range to instead descend the western side in geologically challenging conditions. It was also undertaken at a time when knowledge about geology, and river flows and behaviour, was much more limited than it is today.[157]

Once completed, the scheme had only modest capacity by today's standards (19,200 kilowatts). However, the power it produced was eagerly sought by communities in the region, which scrambled to establish power boards to set up reticulation networks in time for the scheme's commissioning.[158]

This photograph is one of a series produced by Charles showing the construction of the Mangahao hydroelectric power scheme in the Tararua Range east of Shannon. This view shows the construction camp at the lower dam – the original No.1 dam. The cookhouse and the men's huts are on the left bank, and the multiple-cable suspension bridge can be seen to the right.

C.E. Wildbore, Palmerston North Library

from 1902 – with one retailer in Dunedin and one in Whanganui.[153] Retail sales of these cameras petered out around 1906, and so it is possible that Charles acquired his through a private sale.[154]

A further mystery is how Charles travelled to his photographic subjects, particularly those located further afield (such as the Mangahao power scheme, described below). Charles never drove, nor did the family own a motorised vehicle.[159] (Even if he had learned to drive, the purchase price for a motor car in the early 1900s was prohibitive, costing more than senior public servants earned in a year. As a result, the market was limited initially to professionals, especially doctors, and wealthy sheep farmers.[160]) Instead, Charles relied on a more 'traditional' means of transport – horse and cart. He was often seen riding on his dray (a two-wheeled cart) behind his horse, Old Bluey, 'to collect firewood or carry shingle for road repairs'.[161] But it is not clear whether Old Bluey was also entrusted with the important task of carrying a precious cargo of heavy camera equipment and glass plates over long distances. Nevertheless, somehow Charles managed to transport a cumbersome wooden camera, a tripod and glass plates to the various sites he photographed.

The Raumai Bridge on the Pohangina River, damaged by a slip in August 1907. The bridge had opened just the previous year.

Palmerston North Library

Between 1922 and 1924, when he was in his early 60s, Charles produced a series of photographs of the Mangahao power scheme, in the Tararua Range east of Shannon, as it was being constructed (see pp. 62-3).[162] It appears that Charles used a half-plate camera for this series of photographs.[163]

How Charles got to Mangahao is unclear. It is unlikely he attempted the long journey with his horse, Old Bluey. It is possible that a relative or friend took him as far as Shannon by motor car (or he took a train), and from there he hitched a ride on a work lorry up the very rough track to the site.[164]

It is probable that Charles submitted some of his Mangahao photographs to newspapers for publication. His photograph of the powerhouse under construction appeared in a special issue of the *Evening Post* in October 1923, entitled 'The coming of electricity: Mangahao and its promise'.[165]

A dawning 'environmental' awareness?

While Charles may not have been fully aware of the implications of the wide-scale environmental transformation at the time, he was photographing evidence of it. An example of this is a photograph of the aftermath of the slip that seriously damaged the newly erected Raumai Bridge in August 1907.[166] Pohangina Valley residents had been lobbying for the construction of a bridge over the Pohangina River, to allow them to safely traverse the river even in times of flood.[167] Finally, a bridge was completed in 1906, and the official opening ceremony was attended by none other than Sir Joseph Ward, the premier of New Zealand. Ward

was presented with a letter by prominent settlers, which expressed their gratitude for the bridge after more than 10 years of 'protracted delays and disappointments'. The letter further stated:

> *We further desire to express our indebtedness to the Government for the generous manner in which it has contributed to the cost of the structure, the importance of which to the settlers on the other side of the river it would be difficult to estimate. Ever since the settlement of the district the crossing of the river in flood time has been a source of great anxiety to the residents, and we feel that the removal of this disability will more than compensate for the expense involved in the construction of the bridge.*[168]

However, amid the celebrations surrounding the bridge's long-awaited opening, there were those who harboured grave doubts about the approach to the bridge from the Awahou side of the river, where the road had been cut into a steep cliff. Their misgivings turned out to be justified: in the winter of the following year, a period of persistent rain undermined the cliff, and brought down a large slip of earth and rock, completely blocking the approach road on the eastern side. Soon after, a further slip engulfed the first span of the bridge itself, causing it to collapse.[169]

Faced with this major setback, and as an interim measure, Pohangina County Council decided to replace the damaged span of the bridge with a suspension bridge to allow drovers and people on foot or bicycle to get across the river. However, in July 1908, while this structure was still under construction, it, too, was carried away by a flood, and the approach road was again completely submerged by a slip.[170] This meant that the only means of getting across to the eastern side of the Pohangina Valley was to ford the river, or, in times of flood, to cross it using a newly installed cable and cage at Piripiri.[171]

Below: A glass-plate negative of one of Charles' photographs, dated circa 1910 (held by the Palmerston North Library). This image is thought to be of the Pohangina River, looking south.

David Lupton

Above: A "positive" print of the glass-plate negative on the left. The Ruahine Range can be seen in the distance. Note the slip in the face of the cliff, centre.

C.E. Wildbore, Palmerston North Library

Over subsequent decades, the bridge or its approach road from Awahou was periodically damaged or blocked by slips or floods, including in August 1921, when the bridge was 'partially wrecked' by a big slip from the cliff,[172] and in August 1939, when a slip again blocked the road.[173] The tribulations with the bridge were not to cease until it was replaced entirely in 1973 by a new concrete structure, which spanned the river just north of the existing bridge, with the road realigned to avoid the troublesome cliff below Awahou South Road entirely. (This proved to be a wise decision, as the cliff continues to slip today, in the same location the bridge once spanned.)

The Raumai Bridge slip may have alerted Charles to the precariousness of much of the Pohangina District's hill country, now stripped of its vegetative cover. The land's vulnerability to slips stems from its 'geological youth'. Until relatively recently (in geological time), the Pohangina Valley – like

all of the Manawatū Basin – was under the ocean. Much of the underlying rock in the area is comprised of marine sand, similar to that found on New Zealand's west-coast beaches today, which was left behind when the ocean retreated hundreds of thousands of years ago.[174] Since then, the hill country has formed as the land has 'folded' and thrust upwards; conversely, streams and rivers have rapidly cut down through these soft marine sediments, creating unstable cliffs and slopes.[175] Owing to the nature of its geology – underlying rock characterised by unconsolidated sandstone and siltstone, and covered by shallow soils with a weakly developed structure that is easily infiltrated by water – land in the valley is highly susceptible to erosion.[176] For a time, the roots of the incinerated or amputated trees continued to hold the soil together, but eventually they, too, decomposed, leaving only pasture to keep the topsoil in place.

The cyclone that swept through the North Island in early February 1936, just a year before Charles died,[177] is likely to have left the valley battered, and scarred by large slips. But he may well have been dismayed to learn that what he saw in his lifetime would be mild in comparison to the damage wreaked by storms in decades to come. Most notably, Cyclone Bola in 1988 and the North Island storm of 2004 left the valley ravaged by scars, and monumental quantities of topsoil slipped down hillsides, with much of it eventually lost into waterways and out to sea.

Slip on the Pohangina River. Pohangina Valley historian Jim Lundy identified this photograph as a view after the floods of 1907–08, and taken from the bluff opposite Pohangina Village, looking downstream (south) towards Raumai.

C.E. Wildbore,
Te Manawa Museum

Only one area of forest – what today is called Totara Reserve – escaped the wholesale destruction of forest that paved the way for agriculture in the valley (see inset, pages 70 –71). This state forest reserve was the subject of many of Charles' photographs, but the series showing a car – diminutive compared to the trees towering above it – travelling down a narrow metal road through the reserve perhaps leaves the greatest impression. Thankfully, we are able to experience the sense of awe that the car's passengers must surely have felt, as this area of lowland forest still exists today, even though most of the tōtara giants for which it is named fell to the saw long ago.

Charles Wildbore demonstrated remarkable foresight and intuition in capturing everyday events during a time of unprecedented environmental change; events preserved in photographs that are so appreciated today as a window into the past. Sadly, however, Charles' foresight did not extend to the preservation of his photographic legacy. In 1931, six years before his death, he paid two local boys to clean the images from a large pile of glass plates, so that he could use the plates as cloches for his vegetable garden.[178] Despite this loss to 'upcycling', a substantial number of glass-plate negatives survived into the 1940s (mainly of family and other group portraits), when Charles' grandchildren would play with them. Today it is likely that about 100 glass plate negatives exist, only four of these in a public collection.[179]

TOTARA RESERVE

Car in Totara Reserve, circa 1916–20. The vehicle has been identified as a Maxwell, possibly a 1916 Model 25 Touring. The occupants are unidentified but it is interesting to note that there is a woman in the driver's seat. The United States-based Maxwell Company was one of the first car companies to specifically market to women. In 1909 the company generated a great deal of publicity when it sponsored Alice Huyler Ramsey, an early advocate of women drivers, as the first woman to drive coast-to-coast across the United States. By 1914 the company had strongly aligned itself with the women's rights movement. W.H. Palmer of Palmerston North became an agent for the Maxwell from around 1917 (and there were also agents in Feilding and Shannon), so it is possible that a number of these vehicles were on Manawatū roads.

Palmerston North Library

Ironically, Totara Reserve – nowadays a popular destination for camping, swimming, walking and picnicking, as well as a rare example of lowland forest in the region – only survived to be enjoyed today because of its value as a source of timber. The reserve was originally gazetted in 1886 under the provisions of the State Forests Act 1885 as a 'reserve for growth & preservation of timber and for river conservation purposes'.[180] Such sections of forest were set aside primarily for their timber resources, with the intention that they would eventually be milled. This policy was prompted by the growing realisation that timber was a limited resource. However, it also reflected a mounting concern that deforestation was contributing to droughts and downstream flooding – links that were seen in the increasingly frequent and destructive floods that afflicted the Pohangina Valley following bush clearance.

Nationally, there was a growing impetus to preserve indigenous forest for its scenic value, and this objective was first allowed for by the Land Act 1892. Later (and more specifically) the objective was brought under the Scenery Preservation Act 1903 – legislation which allowed for the preservation of land for its scenic value. The move to preserve forested environments for their scenic value was, in part, also motivated by economic considerations, underpinned by a recognition of the potential of New Zealand's natural scenery to draw tourists. In particular, the government prioritised the preservation of forest remnants adjacent to the major travel routes, such as the North Island's main trunk railway line between Wellington and Auckland. Thomas Donne, the first superintendent of the Tourist and Health Resorts Department, established in 1901, believed that 'magnificent areas of forest ... preserved in its primeval beauty contiguous to the railway, will afford a great attraction to travellers ... whereas miles of burnt and blackened logs would prove a weariness to the spirit'. Donne also saw the value in preserving forests so that they would 'serve as a last home for the rarer New Zealand birds'.[181]

But the welcome effects of the scenery preservation movement came late to the Manawatū region. This may be due, in part, to the fact that there was so little left to preserve, but also because the Manawatū District, and especially the Pohangina Valley, were so removed from the popular tourist routes. It was not until 1932 that a portion of Totara Reserve – where the original camping ground and picnic area were later located – was designated as a scenic reserve under the provisions of the Scenery Preservation Act, with its administration vested in Pohangina County Council. Later, in 1947, the remainder of the reserve – until then designated as a state forest reserve – became a public domain, reflecting the recognition of both the scenic and recreational values of the reserve.[182] However, by that time, most of the large tōtara trees in the reserve had been milled. Today, only their stumps remain, serving as a poignant reminder not only of how much the landscape has been transformed in little more than a century, but also of how much our attitudes towards our indigenous heritage have evolved.

Part 4

LEARNING FROM THE LANDSCAPE

People are only able to live and have livelihoods in the Pohangina Valley today thanks to the backbreaking work undertaken by its pioneering settlers, who endured harsh and uncomfortable conditions in their efforts to carve out a livelihood from the land. But we are also continuing to live with the legacy of this dramatic environmental transformation; erosion, floods and deteriorating water quality are all products of actions taken in the past. At the same time, we are witnessing a new era of environmental transformation. Where land was stripped bare of its forest, tōtara, mātai, rimu and rewarewa are again setting down roots and reaching their limbs to the sky, providing places for the berry-eating kererū (wood pigeon) to alight noisily to rest and confer with its companion before again taking flight.

Pohangina wetlands, restored by Gordon and Anne Pilone since 2000.

Catherine Knight

OHANGINA WETLANDS
OPEN

And just outside the Pohangina Village, where Charles himself once farmed, a wetland is being restored on land whose swampy origins have haunted and obstructed successive owners ever since it was first cleared. Today, species of trees and plants well-suited to wetland environments are springing up under the watchful canopy of a scattering of tall kahikatea – survivors of the milling and burning of a century before. In addition to the more common (and resilient) indigenous species, such as pūkeko and paradise shelduck, rarer species, such as tētē (grey teal), kōtuku (white heron), white-faced heron and royal spoonbill, have made a return to this wetland oasis in the valley.[183]

It would be intriguing to know what Charles Wildbore would think of his beloved Pohangina Valley if he could see it today. It is impossible to know whether he would celebrate the slow return of some of the forest trees

Carting railway sleepers on Opawe Road, circa 1904. The sleepers were cut from state-owned land (to the right of the road), and the wagon is about to transport them to the railway station at Ashhurst. The men cutting the sleepers lived in the tents visible in the background.

C.E. Wildbore, Palmerston North Library

The same view on Opawe Road today. Note that the alignment of the road is slightly different.

Catherine Knight

and birds once banished by fire, axe and saw, or regret this regeneration as undoing the hard-won accomplishments of the pioneering bush settlers he photographed.

Whatever his response might have been, we can be grateful for his sharp eye, dedication and intuition. Without his photographic legacy, we would have much poorer insight into the environment of the past, and the potential for transformative environmental change – both negative and positive – at human hands.

GALLERY

Carting sawn timber from Wilson Brothers' sawmill, operating in the vicinity of Totara Reserve, circa 1900. The horses and wagons in this photograph were owned by Pat Hanlon of Ashhurst. This photograph is thought to have been taken on the eastern side of the Pohangina River in the vicinity of Totara Reserve. The road in the photograph may be the old road which went through Totara Reserve (south of Opawe Road) along the riverbank and crossed over to the west side of the river about three kilometres north of Beehive Creek.

C.E. Wildbore, Palmerston North Library

Palmers General Store at Raumai, circa 1910–15. It closed only a few years later, in 1918. The store owner at the time, Mr H.I. Christensen, attributed its closure to the advent of the motor car.

C.E. Wildbore, Palmerston North Library

Sheep farm, circa 1900–10. The exact location of this farm has not been established, but it is thought to be in the Pohangina Valley. The hills in the background are probably the Ruahine Range, and the view seems to be looking south towards the Manawatū Gorge. A river can be seen in the middle distance.

C.E. Wildbore, Palmerston North Library

Sales yards at Pohangina township, circa 1900–10. The yards opened in January 1895, and had a capacity of 2000 sheep and 100 cattle. They were situated about 200 metres south of the township.

C.E. Wildbore, Palmerston North Library

The original, unmodified photograph of the 'travelling tinker'. See also the variations of this photograph on page 57.

C.E. Wildbore.
Lindsay Family collection.

The Wildbore children on the farm in Pohangina, circa 1902. From left to right: Owen, Ruby, Stan, Kathleen, Sylvia(? in front), Charlie, and Percy. The tall plants around the fenceline are tobacco, with pumpkins and possibly onions in the foreground.

C.E. Wildbore, Lindsay Family Collection, Palmerston North Library.

Jane Wildbore with her daughters Ruby Sandbrook (left) and Sylvia Bowditch (right) and granddaughters (front, from left to right): Gail Bowditch, and Norma and Jean Sandbrook, circa 1920.

C.E. Wildbore, Lindsay Family Collection, Palmerston North Library.

Packhorse and whare in the Pohangina Valley, possibly near Utuwai, circa 1900–1910. The whare is typical of the temporary homes built by bushwhackers, grass-cutters and other seasonal labourers – constructed with slab walls and chimney, corrugated iron roof and sack door.

C.E. Wildbore, Palmerston North Library.

A train hauling felled trees in the Pohangina Valley.

C.E. Wildbore, Lindsay Family Collection, Palmerston North Library.

Men sawing timber for railway sleepers, circa 1904. The sleepers were cut from Crown land (part of the Totara Reserve), in bush that had been partly damaged by fire two years earlier. The men received 4 shillings and sixpence per sleeper, delivered to Ashhurst Railway Station, where they were inspected by an official from the Railway Department. The photograph was taken about 300 metres southwest of Opawe Road, about 1.5 km east of the junction with Pohangina Valley East Road. Jack Enright is seen far right with squaring axe. Brothers Ernest and George Wagstaff are third and fourth from right respectively.

C.E. Wildbore, Palmerston North Library.

Cutting railway sleepers near Opawe Road, circa 1904.

C.E. Wildbore, Palmerston North Library.

Threshing grass for seed. Men are seen harvesting the grass on the hill in the background. Behind the threshers is a pile of straw (grass that has been threshed), while the sacks contain the threshed seed. The man standing on the fallen tree is carrying a bundle of cut grass to the threshers. This is likely the same location as the photograph on page 19.

C.E. Wildbore, Palmerston North Library.

Sledging milk from a cowshed, Pohangina, circa 1910.

C.E. Wildbore, Palmerston North Library.

The Pohangina general store, which was situated next to the Methodist Church on Pohangina Road. Taken between 1890 and 1910.

C.E. Wildbore, Palmerston North Library.

WILDBORE FAMILY TREE

Charles Lee Wildbore
1833–1916

Emma Lydia Guy
1837–1916

Thomas Rought Dallison
1834–1921

Jane Wilson
1838–1922

Charles Edward Wildbore
1861–1937

Harriet Eliza Wildbore
1864–1864

John Lee Wildbore
1865–1938

Harriet Eliza Wildbore
1868–19??

Emma Susan Wildbore
1870–1873

Willie Wildbore
1876–1876

Frederick Lee Wildbore
1877–1934

Louisa (Maud) Wildbore
1881–1926

William Alfred Wildbore
1886–1973

Thomas Rought Dallison
1867–1872

Emily Jane Dallison
1869–1942

Mary Eliza Dallison
1870–1935

William Henry Dallison
1874–1938

Celia Dallison
1876–1956

Charles Henry Dallison Wildbore
1889–1944

Stanley Edward Wildbore
1890–1967

Kathleen Germaine Wildbore
1892–1922

Ruby May Cecilia Wildbore
1894–1968

Percy Dallison Wildbore
1895–1967

Owen Dallison Wildbore
1897–1983

Mana Huia Wildbore
1898–1970

Sylvia Jane Wildbore
1899–1970

Baden Hector Archibald Wildbore
1901–1973

Ivan Cecil Wildbore
1903–1971

Evangeline Grace Wildbore
1904–1982

Reginald Mervin Wildbore
1905–1941

Thomas Norman Wildbore
1910–1913

MAP OF LOCATIONS

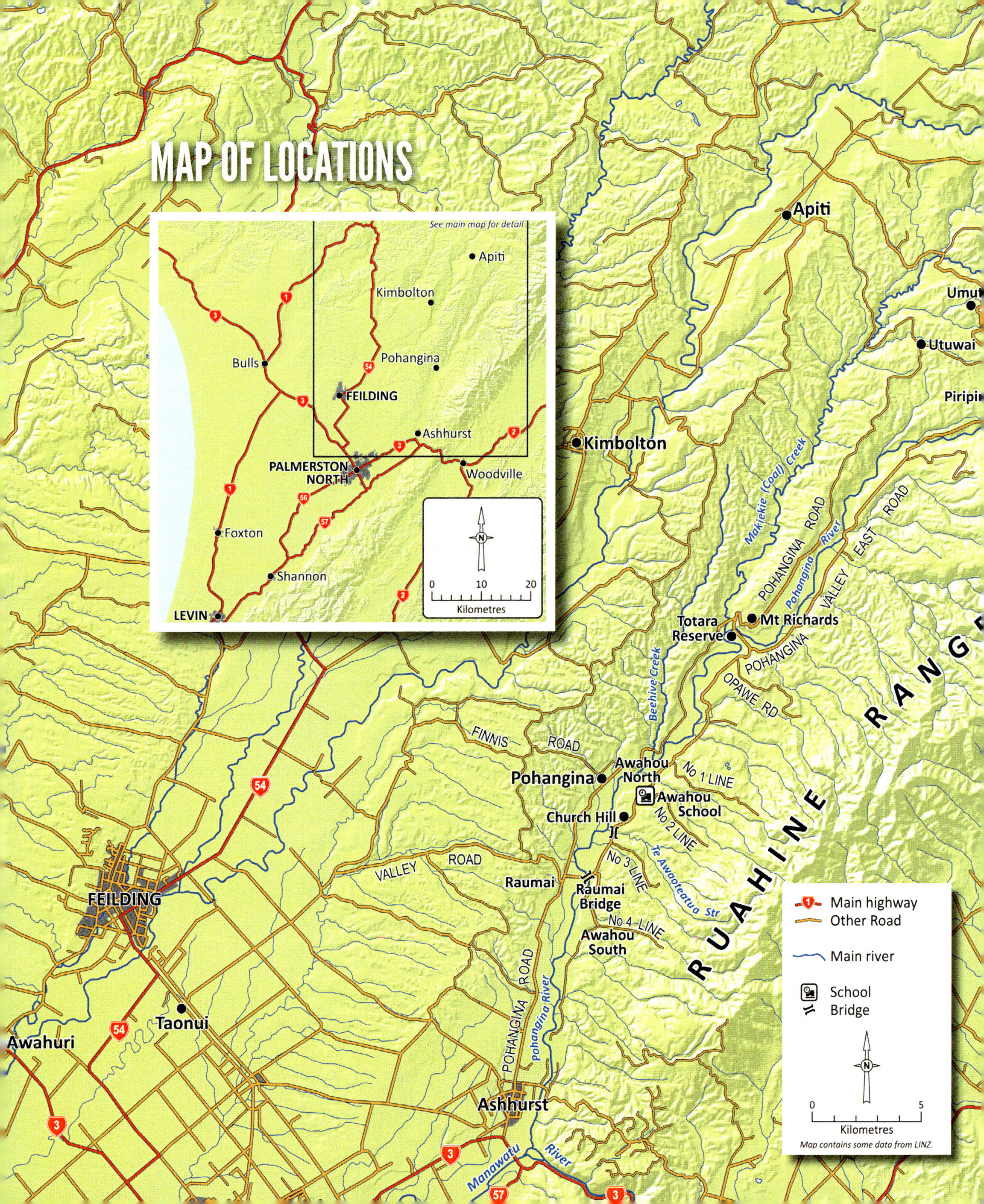

PROVISIONAL PLAN OF POHANGINA COUNTY

circa 1902–06

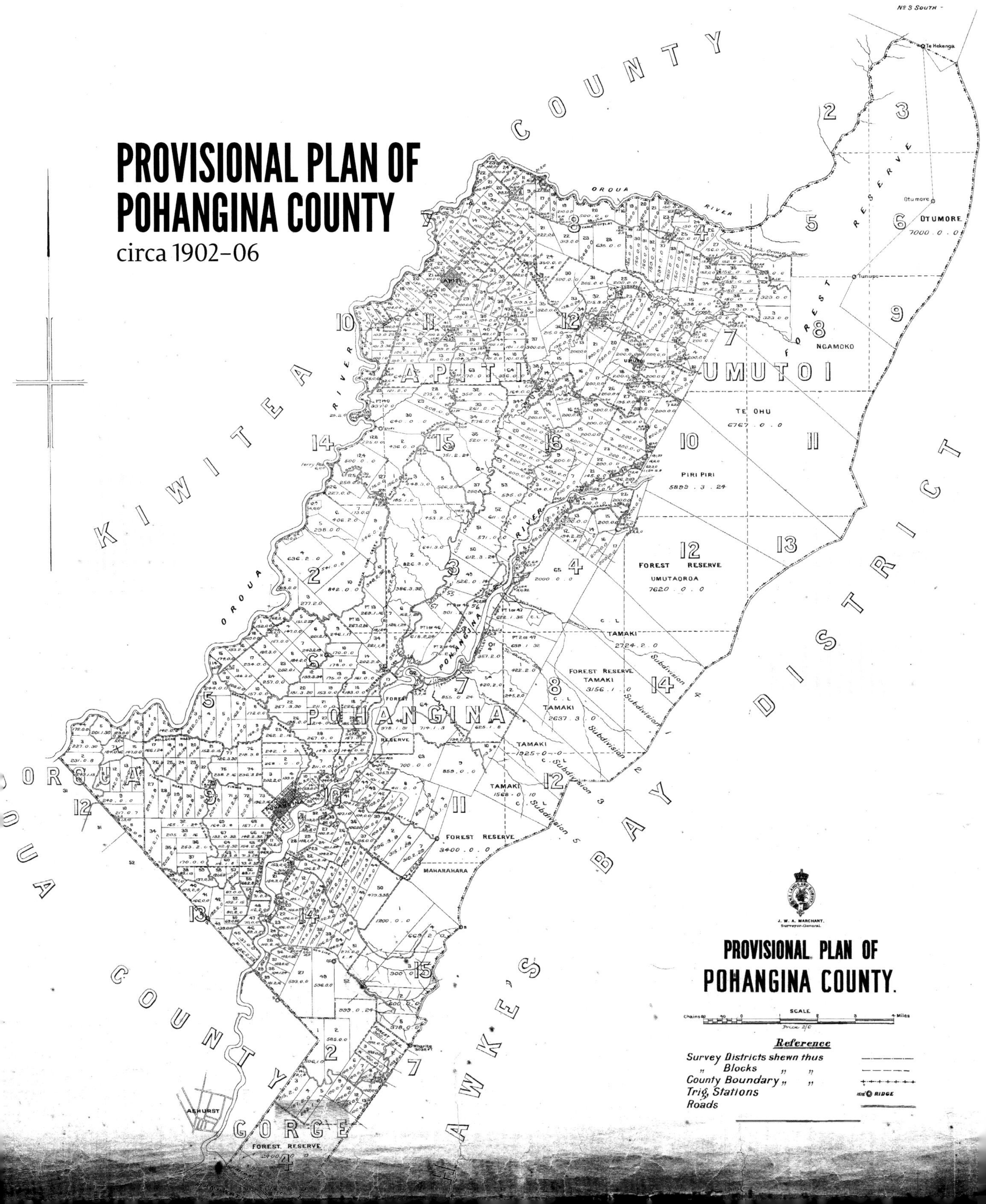

ACKNOWLEDGEMENTS

I would like to thank the following for making this book possible. First and foremost, the Earle Creativity Trust, and its founders Mary and Richard Earle, for supporting this book through their generous grant. Thanks are due, too, to Professor Michael Roche for suggesting this research and urging me to apply to the Earle Creativity Trust.

I would like to thank the people who have helped me with my research along the way: Michael Biggs, senior archivist at Archives Central, Cindy Lilburn at Te Manawa Museum, Heather Glasgow and Leslie Courtney at Palmerston North City Library, Sophie Willis and Anthony Dreaver. A special thanks to Roger Williams for so generously sharing with me his meticulous genealogical research of the Wildbore family, and to Graeme and Jean Lindsay for sharing their collection and knowledge of Wildbore's photographs with me.

My heartfelt thanks to Athol McCredie, Curator of Photography at Te Papa Museum, for sharing his expertise with me and for reviewing the section of the manuscript relating to photography. I am grateful, too, to Bettina Anderson and David Lupton for their input on early drafts of the manuscript and their enthusiasm and support for the project.

I am grateful too for the painstaking research of earlier historians and researchers who prepared the road for me to write this book: Jim Lundy, John Brock and, of course, Ian Matheson among them. These individuals undertook their research at a time when access to material and information was much more limited (and certainly not at one's fingertips) than it often is now. Their achievements cannot go without acknowledgment.

Thank you, too, to Helen Eastwood for her meticulous editing of the manuscript, and to Anthony Behrens for his layout and design of the book. Thanks are due also to Tessa Whiteman for her careful proofreading.

Last of all, I would like to express my appreciation to my husband and business partner, Clive Madge, for his unquestioned support for this project.

NOTES

PART 1

1 A map of locations referred to in this book is provided on page 82.
2 For example, see Pawson & Brooking, 2002; Sinclair & Harrex, 1978.
3 Wright, 1968, p. 62.
4 These areas had originally been forested, but the forest had been destroyed – most likely by fire – before the European colonisation of New Zealand. When Europeans began settling New Zealand in the 19th century, these regions were mostly covered in scrub or tussock, although there were surviving areas of forest in valleys and swampy places that successive fires had not destroyed. Because these regions did not need to be cleared of dense forest, and in some cases could be grazed without any further effort, they were – unsurprisingly – attractive to the early settler.
5 It also created a class of small grazing runs of up to 5000 acres (2023 hectares) that could be leased by smaller sheep farmers. Rent was only 2.5 per cent of the value and, since the land was often rough bush country worth only about five shillings an acre, farmers were charged extremely low rents so they could put their money into stock and improvements (J. McAloon, 2008, 'Land ownership – Centralisation after 1870', *Te Ara – The Encyclopedia of New Zealand*, accessed from https://teara.govt.nz).
6 H. Stringleman & R. Peden, 2015, 'Sheep farming – The refrigerated meat trade', *Te Ara – The Encyclopedia of New Zealand*, accessed from https://teara.govt.nz.
7 Influenced by a landmark Supreme Court ruling in the mid-1880s, the government viewed all land not cultivated by Māori as being 'waste land', and thus reverting to the Crown. Governor George Grey encouraged mass purchase of Māori land, and Māori were advised to gain Crown title (J. McAloon, 2008, 'Land ownership – Māori and land ownership', *Te Ara – The Encyclopedia of New Zealand*, accessed from https://teara.govt.nz).
8 Vogel's ambitious public works and assisted-immigration programmes were initiated in 1870, when Vogel was colonial treasurer, with William Fox as premier. Vogel himself was premier from 1873 to 1875, and again in 1876.
9 It should be emphasised that the archaeological record indicates several centuries of Māori settlement history in the Manawatū region, particularly in coastal areas. That history is not covered here, but is explored in detail in *Ravaged Beauty: An environmental history of the Manawatu* (Knight, 2014).
10 Knight, 2014, pp. 53–79.
11 Tramlines were temporary rail lines used to transport timber and other materials.
12 See Knight, 2014, p. 88 (map).
13 *Manawatu Times*, 19 March 1879.
14 The cost of transport to the railway at Ashhurst was prohibitive, so most of the milled timber was likely to have been utilised to meet local requirements (Knight, 2008).
15 *Feilding Star*, 9 September 1893. The sawmill was evidently still in operation three years later, as the owners attended the Timber Conference held in Wellington in July 1896 (*Appendix to the Journals of the House of Representatives*, 1896, accessed from https://paperspast.natlib.govt.nz/parliamentary/AJHR1896-I.2.3.4.31).
16 Holcroft, 1977, p. 40. Holcroft notes how the value of wages was dependent on whether or not the labourer had a horse and cart, with the maximum being earned by those who had two horses and a cart.
17 Arnold, 1994.
18 Wright, 1966, p. 71.
19 The *Feilding Star* (29 December 1897) reported that a number of settlers in Apiti had undertaken bush burns in the summer of 1897, for example.
20 *Feilding Star*, 23 December 1890.
21 In 1896 Charles Wildbore himself applied to the council to cut grass seed from Finnis Track (evidently not yet a road) and Pohangi-

na Road (*Feilding Star*, 14 December 1896).

22 Matheson, 1971, pp. 52–53; Waugh, 1986, p. 15.

23 Knight, 2013, p. 105.

24 *Timaru Herald*, 4 September 1891. The weather conditions required for a successful burn were quite specific, which meant that, in some cases, two or more years would pass before a farmer could attempt a burn.

25 *Manawatu Times*, 19 March 1879.

26 *Feilding Star*, 18 February 1892.

27 Generally, only the stumps of the larger trees remained intact, and had to be removed before the paddock could be ploughed (Matheson, 1971, pp. 54–55).

PART 2

28 Charles Lee Wildbore was born in 1833 in Kirton in Holland, Lincolnshire, England.

29 Roger Williams (grandson of Eva Wildbore, and great-grandson of Charles Edward Wildbore), pers. comm., 8 April 2018.

30 He was described as a 'journeyman farrier' on the occasion of the 1861 census, when he was 28 years old. 'Journeyman' referred to anyone who had finished learning a trade and was working for someone else rather than for themselves.

31 Emma Guy was born in 1837 in Middlesex, London.

32 Intriguingly, Charles' birthplace has a strong connection with cameras and photography, although he would not have known that when he lived there. In about 1902, the Kodak United Company opened a factory and storefront at 41–43 Clerkenwell, and took advantage of the surplus of unemployed watchmakers to build and repair their cameras.

33 'Cross's new plan of London 1861', accessed from http://mapco.net/cross1861/cross.htm.

34 A. Baggs, D. Bolton & P. Croot, 1985, 'Islington: Growth', in T. Baker & C. Elrington (eds.), 'A history of the county of Middlesex: Volume 8, Islington and Stoke Newington parishes', pp. 9–19. *British History Online*, accessed from www.british-history.ac.uk.

35 *Islington Life* (online magazine published by Islington Council, London), accessed from http://islingtonlife.london/discover-islington/local-history/clerkenwell.

36 Atlantic Publishing's edition of Charles Dickens' *Oliver Twist*, 2014, p. 47; *Islington Life* (online magazine published by Islington Council, London), accessed from http://islingtonlife.london/discover-islington/local-history/clerkenwell.

37 The 1871 census shows that residents were predominantly from working-class occupations, such as farriers, wheelwrights, charwomen (cleaners), servants and labourers.

38 The geographical orbits of Charles Wildbore and Charles Dickens also overlapped in nearby Kent. The Wildbores spent some years in the 1860s in Strood, near Rochester, only a few kilometres from Dickens' summer house at Gads Hill, Higham. However, the story always referred to the two meeting in London rather than in Kent.

39 *Oliver Twist or The Parish Boy's Progress* (Dickens, 2014, chapter X).

40 *Oliver Twist or The Parish Boy's Progress* (Dickens, 2014, chapter VIII).

41 Fortunately it was also the last, as it occurred just as the construction of London's sewage and water treatment network was being completed. At the time of the epidemic, the section in the East End was yet to be finished.

42 In the 1840s and 1860s, conflict over sovereignty and land led to a series of battles between government forces and some Māori tribes.

43 J. Phillips, 2015, 'History of immigration – The great migration: 1871 to 1885', *Te Ara – The Encyclopedia of New Zealand*, accessed from https://teara.govt.nz.

44 Ibid.

45 In total, around 44,000 people migrated to New Zealand in 1874, but approximately 6000 of these subsequently emigrated, leaving a net migration to New Zealand of 38,000 people. This was a record number, which was not repeated until 2002 (J. Phillips, 2015, 'History of immigration – The great migration: 1871 to 1885', *Te Ara – The*

Encyclopedia of New Zealand, accessed from https://teara.govt.nz). In his monograph on the subject of emigration from rural England to New Zealand, Rollo Arnold refers to 1874 as the year of the 'flood tide' of immigration (Arnold, 1981).

46 *SS Euterpe* was a full-rigged iron ship built in 1863 by Gibson, McDonald & Arnold in Ramsey, Isle of Man, and was renamed *Star of India* in 1906 (Moore–Brickland family history, accessed from www.moore.org.nz/genealogy).

47 From Great Britain, ships sailed south through the Atlantic Ocean to utilise the 'roaring forties', the powerful westerly winds found at 40 degrees latitude. From there, the ship sailed due east through the southern Indian Ocean, before entering the southern Pacific Ocean.

48 Moore–Brickland family history. The name 'Whanganui' was officially gazetted in November 2017 although it is likely that in Charles' time the name was commonly spelled as 'Wanganui'. The different spellings arose from the way in which local Māori pronounced the word 'Whanganui' (with a lightly aspirated 'wh') and the way European settlers heard and recorded it (as 'Wanganui').

49 *Wanganui Chronicle*, 19 February 1875.

50 G. Price, 1930, 'International order of good templars', in J.M. Murray & J. Cocker (eds.), *Temperance and Prohibition in New Zealand*. London: The Epworth Press, p. 192, accessed from http://nzetc.victoria.ac.nz.

51 *Wanganui Herald*, 23 October 1876.

52 *Feilding Star*, 26 March 1885; *Feilding Star*, 4 December 1888.

53 *Manawatu Times*, 10 April 1878.

54 *Wanganui Herald*, 10 April 1878.

55 *Wanganui Herald*, 22 May 1878; *Manawatu Herald*, 24 January 1879.

56 See *Feilding Star*, 2 May 1891, for a revealing meeting of creditors in the estate of Charles Lee Wildbore (the proceedings suggest that Wildbore had gone to some lengths to vest assets in other members of the family so that there was little for creditors to lay claim to), and *Feilding Star*, 8 October 1892, which features his appearance at the bankruptcy court.

57 Advertisement from the *Manawatu Times*, 17 January 1880. Foxton Line was renamed Pioneer Highway in 1955.

58 *Feilding Star*, 16 October 1886.

59 *Feilding Star*, 4 November 1886.

60 *Feilding Star*, 3 March 1888.

61 *Feilding Star*, 17 September 1887; *Feilding Star*, 20 September 1887.

62 Newspaper reportage of the time appears to support Wildbore's theory, at least to the extent that there was no obvious (accidental) cause for the fire. According to the *Feilding Star* (20 September 1887): 'The cottage which was so mysteriously burned down on Friday morning, at Taonui, was only occupied at night by a young man employed by the owner, Mr Wildbore. There had been no fire in the place for two days. There were no loose matches about, and the candle used by the man to dress himself before going out to a dance out [*sic*] that night, was brought of [*sic*] the place by a son of Mr Bailey ...'.

63 *Feilding Star*, 14 April 1887.

64 *Wanganui Herald*, 10 October 1887; *Wanganui Chronicle*, 11 October 1887.

65 *Evening Post*, 26 October 1887.

66 *Feilding Star*, 1 November 1887.

67 *Feilding Star*, 10 March 1888.

68 *Feilding Star*, 23 November 1916.

69 Display notes at Pohangina County Fayre, a community-run tearoom and museum housed in the old Pohangina County Council offices. The display notes accompany Charles Wildbore's photographs, and are based on research undertaken by local historian, Jim Lundy.

70 *Manawatu Times*, 15 March 1879.

71 *Feilding Star*, 16 July 1885. He remained a member for a number of years – at least until 1889 (see *Feilding Star*, 27 June 1889; *Feilding Star*, 19 September 1889). At the September 1889 meeting Charles was elected auditor, which suggests he was viewed as a trustworthy character.

72 J. Phillips, 2017, 'Men's clubs – Friendly societies and other fraternal organisations',

Te Ara – The Encyclopedia of New Zealand, accessed from https://teara.govt.nz.

73 There is some confusion about Jane's name. Her birth record identifies her name as Jane, but her will, gravestone and most electoral rolls name her as Emily Jane. Her marriage certificate identifies her as Jane Emily (Roger Wilson, pers. comm., 8 April 2018). She is identified as 'E.J. Wildbore' in the rate books.

74 Her father was from Derbyshire, England.

75 *Manawatu Times*, 30 May 1921.

76 Moore-Brickland family history.

77 Lundy, 1996; Charles Edward Wildbore/ Emily Jane Dallison Family tree at https:// whakapapa.gen.nz.

78 Lundy, 1996; Display notes at Pohangina County Fayre; Roger Williams, pers. comm., 19 October 2018.

79 Display notes at Pohangina County Fayre.

PART 3

80 J.T. Stewart, 'Reminiscences' (undated manuscripts), cited in Petersen, 1952, p. 24. Petersen, 1973, pp. 187, 190. A local newspaper report of 1883 describes it as a 'triangular fern clearing' (*Wanganui Herald*, 17 July 1883).

81 Lange, 2000, p. 9. Note that there is no archaeological evidence of any permanent Māori settlements in the Pohangina Valley itself. There was a Rangitāne pā (fortified village) at Otangaki, at the confluence of the Pohangina and Manawatū rivers, where the Ashhurst Domain is now located (Petersen, 1973, p. 13; Buick, 1903, pp. 32–33).

82 The name was originally spelled with one 'h', as 'Ashurst'.

83 The Manchester Block, which included land in both the Ahuatūranga Block and the adjacent Rangitikei-Manawatū Block, encompassed three major townships – Feilding, Halcolme and Ashhurst.

84 Wanganui Harbour Endowment and Borrowing Bill 1877 (*Wanganui Herald*, 27 September 1877). This and surrounding land in the Pohangina Valley were referred to as 'waste lands' in the parlance of the time – an official term that described all land not cultivated by Māori, and hence deemed to be belonging to the Crown (J. McAloon, 2008, 'Land ownership', *Te Ara – The Encyclopedia of New Zealand*, accessed from https://teara.govt.nz).

85 *Manawatu Times*, 27 October 1877.

86 It was later suggested by one correspondent to the *Manawatu Standard* that the government should have made provision for good access roads before making the endowment to the Wanganui Harbour Board (*Manawatu Standard*, 4 December 1883).

87 *Manawatu Times*, 19 March 1879.

88 *Wanganui Herald*, 6 December 1878. In 1879, Mr Dundas, surveyor for Palmerston (North), advised the Wanganui Harbour Board that Pohangina Road was the best road line for opening up the Pohangina Valley. He also advised on the best location for a township (*Wanganui Herald*, 4 April 1879).

89 *Wanganui Herald*, 21 December 1881. It is not clear how many sections sold when the land in these blocks was offered for sale in March 1882 (*Wanganui Chronicle*, 28 February 1882). However, about seven sections sold when land was again offered for sale in December 1882 (*Feilding Star*, 6 December 1882; *Wanganui Chronicle*, 23 December 1882).

90 In 1882, the engineer (Lloyd Hassell) commissioned by the harbour board to survey the road lines again urged the board to construct road lines before attempting to sell any more land in the endowment block (*Wanganui Chronicle*, 9 September 1882).

91 *Wanganui Chronicle*, 8 November 1882. In April 1883, the harbour board resolved not to pay for the construction of roads except where they would open up access to the unsold sections (*Wanganui Chronicle*, 4 April 1883).

92 In November 1887, the Kiwitea Road Board announced its intention to raise a loan to complete Pohangina Road and Finnis Road, and also to construct sections of Valley, Zig Zag and Taonui roads (*Feilding Star*, 8 November 1887).

93 *Feilding Star*, 1 December 1887. The school was originally located near Holley's Hill (according to notes held on Wildbore's photographs at Pohangina County Fayre).

94 *Wanganui Herald*, 5 June 1890. In 1885 the

Wanganui Harbour Board had written to the Waste Lands Board requesting permission to sell the remainder of the endowment block. The latter board decided to consent to the request if one-third of all deferred payment instalments were distributed to local bodies for the purposes of making roads in the block (*Manawatu Standard*, 2 May 1885).

95 *Feilding Star*, 31 October 1891; *Feilding Star*, 5 December 1891. Settlers were evidently frustrated with the lack of progress on roads. In December 1891, the *Woodville Examiner* noted that 'Mr Lucas of the Pohangina Small Farm Association urges that all the Associations now being formed should request Parliament to empower them to borrow under the Government Loans to Local Bodies Act, the loans to be repaid by the returns, so that funds might be immediately provided for road-making' (*Woodville Examiner*, 16 December 1891).

96 *Feilding Star*, 20 September 1892.

97 By the mid-1890s, some 76 associations with over 2500 members had taken up almost 500,000 acres (202,342 hectares) (J. McAloon, 2008, 'Land ownership', *Te Ara – The Encyclopedia of New Zealand*, accessed from https://teara.govt.nz).

98 *Feilding Star*, 25 March 1893. More sections were offered for sale in 1896 (*Wanganui Herald*, 11 July 1896). This followed a petition by existing residents that urged the harbour board to put more sections up for sale (*Wanganui Chronicle*, 8 June 1895).

99 The other blocks balloted for were VII and VIII. According to the provisional plan of Pohangina County (circa 1902–06) (see p. 83), these were located adjacent to each other on either side of the Pohangina River, to the north of the state forest reserve (known today as Totara Reserve).

100 John Wildbore applied to the Wellington Land Board to surrender his Umutoi section (section 33, Block IV), and the application was accepted (*Feilding Star*, 28 June 1895).

101 Charles purchased this section in the Pohangina Village from its previous owner, J. (Joseph) Husband (*Feilding Star*, 10 February 1896). The Pohangina County Council rate book for 1896 shows 'Chas Wildbore' struck off the rate book for his Umutoi address (section 32, Block IV) and added as a ratepayer for section 2, Pohangina Riding. Previous biographical information (Lundy, 1996) stated that the Wildbores purchased their Kiwi Street property in 1893, but this is not supported by the sources.

102 The harbour board intended to subdivide the land into 20 quarter-acre sections but, at the time, the land was 'standing bush', and was not anticipated to be required for several years. Consequently, such a lease arrangement would have suited the board as it meant that, at the least, the board would gain some income from the land (*Wanganui Chronicle*, 12 March 1897).

103 *Manawatu Standard*, 12 October 1914. Subsequent to this he continued to lease other land from Pohangina County Council (*Manawatu Standard*, 13 November 1916).

104 *Manawatu Standard*, 3 April 1909. A later report refers to sections 55 to 59 (*Wanganui Chronicle*, 1 May 1911).

105 It is possible that a portion of these were leased (at least initially), but the rate book covering this year does not provide any information to confirm this.

106 Pohangina County Council rate books for 1901–02 and 1902–03.

107 This assessment is based on a survey of Pohangina County Council rate books between 1896 (when Charles purchased his first property in Pohangina Village) and 1936, the year prior to his death.

108 The latter explanation is quite possible because the 'township' of Pohangina as visualised by the Wanganui Harbour Board planners was never realised. Most of the town sections remained mere lines on the 1891 plan, as did a number of the streets – and some streets, such as Kiwi Street and Kaka Street, eventuated only in attenuated forms.

109 *Feilding Star*, 20 September 1892; *Manawatu Standard*, 7 August 1907; *Manawatu Standard*, 13 July 1901.

110 Display notes at Pohangina County Fayre. Taylor's General Store was established in 1904, originally to sell liquor (Rita, who works at Taylor's Four Square, pers. comm., 9

April 2018).
111 *Manawatu Times*, 25 March 1903 (this article actually refers to C.F. Wildbore but this is most likely a typographical error).
112 Display notes at Pohangina County Fayre.
113 *Manawatu Times*, 10 May 1927.
114 *Feilding Star*, 4 June 1895.
115 *Greymouth Evening Star*, 23 July 1906.
116 *Taranaki Herald*, 24 November 1913.
117 *Manawatu Standard*, 2 March 1904.
118 Graeme Lindsay, pers. comm., 21 June 2018.
119 The dry-plate technique was first developed by R.L. Maddox in 1871 but took some time to become commercialised and supersede wet-plate photography.
120 The wet-plate collodion process was a photographic technique invented by Englishman Frederick Scott Archer in 1851. The process involved adding a soluble iodide to a solution of collodion (cellulose nitrate), and then coating a glass plate with the mixture. In the darkroom the plate was immersed in a solution of silver nitrate to form silver iodide on the plate's surface. The sensitised glass plate, still wet, was exposed in the camera. The resultant image was then developed by pouring a solution of pyrogallic acid over the plate, and was fixed with a strong solution of sodium thiosulfate (later substituted by potassium cyanide). The fixing solution stabilised the image, removing any unexposed silver iodide remaining on the plate, and leaving behind the metallic silver that formed the image. Immediate developing and fixing were necessary because, after the collodion film had dried, it became waterproof, and the developing and fixing solutions could not penetrate it. While complex and somewhat dangerous owing to the chemicals used, the process was valued for the level of detail and clarity it produced in the images (*Encyclopaedia Britannica*, accessed from www.britannica.com).
121 *Waikato Times*, 14 February 1882.
122 The plates were ready-coated in a factory, as opposed to the photographer having to coat the plates before taking each photograph.
123 *Woodville Examiner*, 25 April 1890. The first transparent plastic roll film was developed in 1889, but was made from highly flammable nitrocellulose ('celluloid'), now usually referred to as 'nitrate film'.
124 Display notes at Pohangina County Fayre.
125 Lundy, 1996.
126 *Feilding Star*, 20 December 1892; *Feilding Star*, 17 March 1897; *Feilding Star*, 11 December 1897.
127 *Manawatu Times*, 10 April 1901.
128 *Manawatu Standard*, 16 February 1907.
129 Display notes at Pohangina County Fayre.
130 'A&P' stands for 'agricultural and pastoral'.
131 *Feilding Star*, 31 October 1906; *Bush Advocate*, 31 October 1906.
132 *Manawatu Times*, 30 October 1907.
133 Mana, Charles' fifth-born son (born 1898), is notable for having Māori first and middle names (his middle name was Huia) – the only child in the family to do so. It is unclear what led Charles and Jane to give their son these names, but it may be that giving Māori names to European babies was a minor fashion at the time, most likely as a consequence of the 'Huia Onslow' phenomenon. When the wife of the governor general to New Zealand, William Onslow (in New Zealand from 1889 to 1892), had a baby boy in 1890, the baby was bestowed with a Māori name, Huia. (In fact, 'Huia' was the child's middle name, but it was the name by which he was known all of his life.) Mana Wildbore had a cousin (born 14 years later, in 1912) who was also given these names, and whose twin brother had 'Tui' as a middle name. An additional source of influence may have come from Jane's uncle, George Wilson, who was the last editor of the newspaper *Te Waka Maori* before it ceased publishing in November 1884. One of George's daughters also had 'Mana' as a middle name (Roger Williams, pers. comm., 2 April 2018).
134 A print of this original image is held by Sophie Willis. She came across it, by chance, in a secondhand shop in Feilding. It apparently came from a house-lot in Levin, possibly from a deceased Wildbore descendant (Sophie Willis, pers. comm., 11 April 2018). Another print (along with the original glass plate negative) is held by the Lindsay family.
135 Graeme Lindsay (grandson of Charles' daughter Ruby), pers. comm., 21 June 2018.

Peter Wildbore (Percy Wildbore's son, and grandson of Charles Wildbore) first alerted me to the existence of more than two versions of the image (pers. comm., 23 December 2017).

136 Sophie Willis, pers. comm., 11 April 2018; Graeme Lindsay, pers. comm., 21 June 2018. Charles apparently talked to the cat continuously to encourage her to look up at him for the shot (Graeme Lindsay, pers. comm., 21 June 2018).

137 Roger Williams, pers. comm., 2 and 3 April 2018.

138 Athol McCredie is the author of *New Zealand Photography Collected* (Te Papa Press, 2015).

139 Athol McCredie, pers. comm., 9 April 2018.

140 It is also possible that he obtained some of his equipment by mail order, potentially through specialist magazines, such as the *Sharlands Trade Journal* (produced by the pharmacy industry), which included a photography section from 1892, and a dedicated photography magazine from 1896.

141 *Feilding Star*, 3 January 1891.

142 *Feilding Star*, 10 February 1903.

143 *Auckland Star*, 10 April 1908.

144 For instance, in 1902 the Brown No. 2 camera sold for 12s 6d (12 shillings and sixpence) at Enos Pegler's store in Palmerston North (*Manawatu Standard*, 26 April 1902).

145 For example, it is said that when he bought a stereoscopic camera costing £20, probably around 1910, it caused a 'spirited' argument between Jane and himself, Jane expressing her intense displeasure at what she considered the frivolous use of money on his time-consuming pastime. (Graeme Lindsay, pers. comm., 21 June 2018). To put this expenditure in context, the average weekly income of a skilled labourer at the time was just under £3, so this would have been a sum equivalent to nearly two months of work.

146 It is interesting to note that, for a period, Charles' father also owned a whole-plate Thornton-Pickard camera. The camera was advertised for sale in 1907 (*Feilding Star*, 30 October 1907).

147 Athol McCredie, pers. comm., 5 April 2018.

148 Lundy, 1996; display notes at Pohangina County Fayre.

149 One photograph is of wheelwrights at work, and the other is of a photographer (see p. 53), both held by Palmerston North Library.

150 Pritchard, 2015, p. 27.

151 *Nelson Examiner and New Zealand Chronicle*, 1 January 1862; *Otago Witness*, 4 January 1862; *New Zealander*, 8 January 1862.

152 *Otago Daily Times*, 20 May 1898.

153 *Otago Witness*, 9 April 1902. Stereoscopic cameras were sold by Wilton's Photographic Depot in Whanganui (*Wanganui Herald*, 7 September 1905).

154 There was a trickle of sales of these cameras in the district in subsequent years, including an advertisement by a private seller for a 'half-plate stereoscopic camera, 3 d.d. slides, behind lens, pulley blind, shutter, Aldis lenses, tripod, carrying case. All nearly new. Worth £12, take £8 cash' (*Manawatu Times*, 15 June 1908).

155 Martin, 1991, p. 80. Indeed, the perceived importance and urgency of the scheme by those in the Manawatū region cannot be overemphasised – so much so that a special-purpose 'Mangahao Hydro Electric Power League' was formed at a conference in Palmerston North in June 1918 to agitate for an immediate start to the project. Further, at public meetings, residents harangued the minister of public works to give the project the go-ahead without further delay (Martin, 1991, p. 81).

156 Martin, 1991, p. 80.

157 Martin, 1991, p. 81. See Knight (2016, pp. 128–29) for a discussion of how the scheme was saved from a potentially catastrophic failure by local sheep farmer, amateur geologist and photographer, G.L. Adkin.

158 Martin, 1991, pp. 81, 86.

159 Sophie Willis, pers. comm., 11 April 2018.

160 E. Pawson, 2014, 'Cars and the motor industry – A motorised society', *Te Ara – The Encyclopedia of New Zealand*, accessed from https://teara.govt.nz.

161 Lundy, 1996.

162 Biographical notes by Jim Lundy (1996) suggest that Charles acquired a stereoscopic camera around this time and took a series of stereoscopic views of the scheme. However, in fact, he had acquired a stereoscopic

camera about 10 years earlier, and, if stereoscopic photographs of the dam were taken, no evidence of these can be found in public archives.

163 This is judging from the dimensions of the prints made from the negatives (held by the Lindsay family), which are about 6 inches by 4 ½ inches (16 x 11.5 cm).

164 G.L. Adkin, who took an extensive series of photographs of the site between 1919 and 1926, rode his horse up to the site (Anthony Dreaver (Adkin's biographer), pers. comm., 10 April 2018). However, Adkin lived in Levin, a much shorter distance from Mangahao compared with the distance from Charles' home in Pohangina, and he was also considerably younger (Adkin was in his early 30s when he first began visiting the site).

165 This photograph is held in the Palmerston North City Library archive (reference number 2011P_Mgh1_005855).

166 In fact, we do not know with certainty that this photograph is Wildbore's work, but it has all the hallmarks of his photography: well-balanced composition, attention to detail, technical precision and the presence of people (four men standing on the road above the bridge, along with a dog, and horse and cart). We also know that the bridge was a subject of interest – he took a photograph (held in the Te Manawa collection) of the newly completed bridge just months before it was damaged by the slip.

167 *Manawatu Times*, 5 December 1906.

168 *Manawatu Times*, 5 December 1906. The government contributed £2000 towards the construction of the bridge, with the Palmerston Borough Council contributing £350, the Feilding Borough Council contributing £225 and the Oroua County Council contributing £525 (*Manawatu Standard*, 6 November 1906).

169 *Manawatu Standard*, 28 August 1907.

170 *Manawatu Standard*, 9 July 1908. The contractor later sued Pohangina County Council for the loss of material and tools caused by the slip (*Manawatu Times*, 1 September 1910).

171 *Manawatu Times*, 9 July 1908.

172 *Manawatu Times*, 8 August 1921.

173 *Horowhenua Chronicle*, 21 August 1939.

174 MacPherson, 1985, cited in Miri, 1999, p. 41.

175 Miri, 1999, p. 41.

176 Miri, 1999, pp. 42–43. 'Unconsolidated' means that rock material is geologically young and has not had time to lithify – that is, to form hard rock. While siltstone is generally stable, sandstone is prone to slump erosion, such as when a mass of loosely consolidated rock moves down a hill slope.

177 Charles died in Palmerston North Hospital on 7 September 1937. He is buried at Pohangina Cemetery.

178 Lundy, 1996.

179 The four glass-plate negatives are held by the Palmerston North City Library.

180 Though the reserve was officially administered by the commissioner of Crown lands (Department of Lands and Survey), Pohangina County Council – as the local council – had a supervisory role in its management.

181 From the *Appendix to the Journals of the House of Representatives*, cited in Star & Lochhead, 2013, p. 149.

182 Its administration was vested in Palmerston North City Council. It was subsequently transferred to Manawatū District Council, and, most recently, to Horizons Regional Council.

PART 4

183 'Pohangina wetlands development', accessed from www.pohangina.org.

BIBLIOGRAPHY

The bibliography lists books, journal articles, theses and archival material. All newspaper articles and online sources are referenced in the endnotes. All historical newspaper articles (up to the 1930s) are accessed through *Papers Past*, https://paperspast.natlib.govt.nz.

Arnold, R., 1981. *The Farthest Promised Land: English villagers, New Zealand immigrants of the 1870s.* Wellington: Victoria University Press with Price Milburn.

Arnold, R., 1994. *New Zealand's Burning: The settlers' world in the mid 1880s.* Wellington: Victoria University Press.

Buick, T.L., 1903. *Old Manawatu.* Palmerston North: Buick & Young.

Dickens, C., 2014. *Oliver Twist.* Hertfordshire, UK: Atlantic Publishing.

Frean, J. & Ravine, D.A., 1986. *An Archaeological and Historical Survey of the Pohangina River Valley, Manawatu, New Zealand.* Palmerston North: Manawatu Museum.

Hall, L. 1995. *Awahou School Centenary 1894–1994: A history of the Pohangina Valley.* Awahou: Awahou School Centenary Committee.

Holcroft, M.H., 1977. *The Line of the Road: A history of Manawatu Country, 1876–1976.* Palmerston North: Manawatu County Council.

Knight, C., 2008. 'Totara Reserve: A window into Manawatu's environmental history'. *The Manawatu Journal of History*, (4), pp. 50–58.

Knight, C., 2013. Creating a pastoral world through fire: The case of the Manawatū, 1870–1910. *The Journal of New Zealand Studies*, 16, pp. 100–120.

Knight, C., 2014. *Ravaged Beauty: An environmental history of the Manawatu.* Auckland: Dunmore Press.

Knight, C., 2016. *New Zealand's Rivers: An environmental history.* Christchurch: Canterbury University Press.

Lange, R., 2000. *The social impact of colonisation and land loss on the iwi of the Rangitikei, Manawatu and Horowhenua region, 1840–1960.* An unpublished report prepared for the Crown Forestry Rental Trust.

Lundy, J., 1996 (October). 'Biographical notes about Charles Wildbore'. Available from the Palmerston North City Archives.

Martin, J.E., 1991. *People, Politics & Power Stations: Electric power generation in New Zealand 1880–1990.* Wellington: Bridget Williams Books.

Matheson, I., 1971. 'The birth of Palmerston North', *Manawatu Evening Standard* centenary supplement, 13 March.

McCredie, A., 2015. *New Zealand Photography Collected.* Wellington: Te Papa Press.

McEwen, J.M., 1986. *Rangitane: A tribal history.* Wellington: Reed Publishing.

Miri, S.D., 1999. 'Erosion and land use in the Pohangina region: A study using GIS and remote sensing'. A thesis presented in partial fulfilment of the requirements for the degree of Master of Applied Science in Soil Science, Massey University, Palmerston North.

Passey, R.G. (Compiled and produced by J. Brock), 1993. *A brief history of Methodism in the Pohangina Valley 1893–1993*. Pohangina: Pohangina Methodist Church.

Pawson, E. & Brooking, T. (eds.), 2002. *Environmental Histories of New Zealand*. Auckland: Oxford University Press.

Petersen, G.C., 1952. *The Pioneering Days of Palmerston North*. Levin: Kerslake, Billens and Humphrey.

Petersen, G.C., 1973. *Palmerston North: A centennial history*. Wellington: A.H. & A.W. Reed.

Pritchard, M., 2015. *A History of Photography in 50 Cameras*. England: Firefly Books.

Sinclair, K. & Harrex, W., 1978. *Looking Back: A photographic history of New Zealand*. Auckland: Oxford University Press.

Star, P. & Lochhead, L., 2013. 'Children of the burnt bush: New Zealanders and the indigenous remnant, 1880–1930', in E. Pawson & T. Brooking (eds.), *Making a New Land: Environmental histories of New Zealand*. Auckland: Oxford University Press.

Waugh, D.T., 1986. *North to Kimbolton*. Feilding: Fisher Print.

Wright, L.A., 1968. 'The axe bites deep: Settlement and land use in the Pohangina County, 1863–1963'. A thesis presented in partial fulfilment of the requirements for the degree of Master of Arts in Geography, Massey University, Palmerston North.